CO
CATA

H Girdlestone

1987

COMET
CATASTROPHE!

ROGER SUTHERLAND

JAVELIN BOOKS
POOLE · NEW YORK · SYDNEY

First published in the UK 1985 by Javelin Books,
Link House, West Street, Poole, Dorset, BH15 1LL

Copyright © 1985 Roger Sutherland

Distributed in the United States by
Sterling Publishing Co., Inc.,
2 Park Avenue, New York, NY 10016

Distributed in Australia by
Capricorn Link (Australia) Pty Ltd,
PO Box 665, Lane Cove, NSW 2066

British Library Cataloguing in Publication Data

Sutherland, Roger
  Comet catastrophe.
  1. Halley's comet.
  I. Title
  523.6'4   QB723.H2

ISBN 0 7137 1843 9

All rights reserved. No part of this book may be reproduced or transmitted in any form or by any means, electronic or mechanical, including photocopying, recording or any information storage and retrieval system, without permission in writing from the publisher.

This book is sold subject to the conditions that it shall not, by way of trade or otherwise, be lent, re-sold, hired out or otherwise circulated without the publisher's prior consent in any form of binding or cover other than that in which it is published and without a similar condition including this condition being imposed on the subsequent purchaser.

Typeset by Furlonger Phototext Ltd, Bournemouth.

Printed and bound in Great Britain by
Cox and Wyman Ltd, Reading Berks.

Cover Photograph : Magnum

CONTENTS

1 Comet Catastrophe 7
2 Harbingers of Doom 11
3 The Star-Crossed Sky 24
4 Past Apparitions 38
5 Earthquakes and Weather 44
6 Blood on the Streets 57
7 Pilot Error? 77
8 A New Plague? 91
9 The Ultimate Catastrophe 100
10 Approaching Perihelion 108

Index 121

1
COMET CATASTROPHE

Anyone who reads the newspapers must have been struck by the fact that the eighteen months or so preceding the closest approach to the sun of Halley's Comet in February 1986 seemed unusually packed with disasters. The record of air crashes alone was terrifying; the same was true of train crashes; there were riots all over the world; the weather was extraordinary, with the hurricanes in America, the Bay of Bengal typhoon and the Australian bush fires, not to mention the harshest winter Europe had seen for decades, and the wettest summer England had seen in almost half a century; there was the Mexico City earthquake; in India, there was the Bhopal chemical plant disaster and the assassination of Indira Gandhi, while in England there was an attempt to blow up the whole of the British government; terrorism generally (especially aircraft hijackings) flourished in a way that it had not done for more than a decade; and the horrific disease of AIDS (Acquired Immune Deficiency

Syndrome) ceased to be a disease which afflicted a few unfortunate homosexuals, and became a source of worry for everyone.

What is all the more remarkable is that the same reading of newspapers discloses that the *first* seven or eight months of 1984 reveals a period of (relative) calm: there does seem to be an extraordinary pattern building up after that. It may of course be pure coincidence (if there is such a thing — we shall investigate Jung's Theory of Synchronicity more thoroughly later in the book), but it may not. It may also be that comets are even more disastrous than we think; our fear of them may be a race memory of the utter destruction of past civilisations, and the death of most of mankind, as well as the wiping from the face of the earth of whole lines of evolution.

Five thousand years, or at most six, separate us from the earliest civilisations of which we have any real knowledge — and mark well that phrase 'of which we have any real knowledge'. The Romans were at their height less than 2000 years ago; the Greeks, less than 2500 years ago; and the Egyptians' Old Kingdom was founded just over 4500 years ago. Roman historians left us a surprisingly complete record, including many translations from the Greek, and the Greeks themselves left quite a lot, but before that most of what we know is either hearsay from Greek and Roman authors, or (if there are any written records at all) either boastful records of battles won or tedious inventories of royal goods and records of merchants' transactions; the last two are about as exciting to read as laundry lists, which they closely resemble.

We do know, though, that most if not all of those civilisations had a highly developed science of astronomy/astrology (to which we shall return in the next chapter), and we know from the writings of the Greeks and the Romans that comets were almost always associated in their minds with disaster. Curiously, although we know a good deal about the astronomy/astrology of earlier races, we do not know much about their attitude to comets: although the Greeks tell us that the Egyptians were afraid of them, and that their own word *kometes* ('hairy') comes from the Egyptian and refers to the 'hairy' tail of the comet, we do not even know the Egyptian hieroglyph denoting 'comet'. Could this be because the Egyptians regarded comets as too important, or too magical, to commit to writing?

In many civilisations, there is a tradition that the most important teachings — the inner secrets — are not written, but are passed on in an oral tradition. In Western civilisation, where everything is written down, this is hard to believe, but the Sufis of Islam, the Vajrayana ('Thunderbolt Path') and Dzogchen ('Great Teaching') Buddhists of Tibet, the Qabalists of Judaism, and many others have stressed the importance of teaching lineages; and, of Britain, Julius Caesar records that the Druids opposed writing, but memorised long and complex histories and rituals. Given that the astronomer/astrologists of many past civilisations were priests, might it not be that their inmost teachings, too, were not committed to paper, papyrus, clay or stone?

For it is possible that comet strikes upon the earth were the cause of those sudden breaks in evolution, and that the words *of which we have any real knowledge* in the phrase 'the earliest civilisations of which we have any real knowledge' are more important than we dare think. Hindu and Buddhist thought refers to the *kalpas* (unutterable aeons) which have passed before, in which civilisations rose and fell, and rose and fell again, in many cycles before our present one; and in Ecclesiastes, Chapter I, we read:

9: *The thing which hath been, it is that which shall be; and that which is done, is that which shall be done: and there is no new thing under the sun.*

10: *Is there any thing whereof it may be said, See, this is new? it hath been already of old time, which was before us.*

11: *There is no remembrance of former things; neither shall there be any remembrance of things that are to come, with those that shall come after.*

The very word 'disaster' derives from the older word *desastre*; *des* meaning 'turned away' and *astre* meaning 'star'; when the guiding stars turn their faces away from us, and abandon the heavens instead to comets, our ancestors believed that terrible times were due. Nowadays, the temptation is to dismiss this belief as empty superstition, as nonsense, as 'old wives' tales'. We all know, though, that modern science has proved that many 'old wives' tales' *do* have some foundation in observable fact; the arrogance of Victorian scientists has given way to a new brand of humility, especially among the leading scientists and thinkers of our time, who recognise how little we really do know.

In the rest of this book, we shall look at the history of comets — and particularly at the history of Edmond Halley, the founder of modern cometology — as well as at the ways in which comets might conceivably affect our lives, causing the disasters which give this book its name, and which are examined in detail in Chapters 5 to 9. At times the history will be dry, as all histories must sometimes be, but at other times there will be theories so wild, and so outrageous, that we almost automatically reject them — though we must always be aware that unless there is a better theory (that is, a theory which explains the facts better) we cannot simply reject them no matter how much we would like to do so.

2
HARBINGERS OF DOOM

Actually, we know very little about comets. It is less than three hundred years since they were fitted into the framework of astronomy, and recognised as (reasonably) predictable heavenly bodies. Unlike most other objects in the sky, the main advances in the study of comets have been made by astronomers, rather than astrologers. This is all the more interesting in view of the fact that they have attracted more popular superstition than any other celestial phenomenon, and because of the way in which they have almost always been regarded as unfavourable omens, though there is also a tradition that they can predict change for the better as well as change for the worse.

What is really intriguing, though, is the way in which the study of comets provides a microcosm of the rise *and fall* of 'classical' science — that is, of the sort of science which was taught in universities until very recently, and which is still taught in schools to this day. Even the most incredible theories, which would have been discarded out of hand by the scientists of fifty or even twenty years ago, are now being studied with interest; and that is what this book is about. Can comets really be harbingers of doom? If so, can we explain how this might be? Are there forces at work which we only half understand — or do not understand at all? In short, can we apply the scientific method itself — the unprejudiced study of facts, as distinct from trying to prove a particular case — to our apparently 'unscientific' subject?

In order to answer this question, we need to go through the history of cometology, and to look at the relationship between astrology and astronomy. The leading name in cometology is of course Halley, the first man to make an accurate prediction of the date of return of a comet, and whose own comet — Halley's Comet — returns to perihelion (its position nearest the sun) on 9th February, 1986. But before we can look at Halley and his comet, we have to go back ten thousand years or more, to the very first astrologers.

Nobody knows when astrology started, but the Babylonians, the Chaldeans, the Sumerians, the ancient Egyptians, and the people of the Indus valley where civilisation started were all familiar with it — and for good reason. It is easy to forget, from our twentieth-century viewpoint, that our familiar calendar is actually based on the movements of the planets (especially our own) and of the moon, but of course the year is the time that it takes the earth to complete a single circuit of the sun, and the word 'month' has the same root as 'moon', reflecting the time it takes for the moon to orbit the earth. To our far-distant ancestors, constant confirmation of the calendar was necessary for calculating sowing, growing, and reaping times, so it is not hard to see how a genuine 'science', which we now call astrology, came to be based on their observations and calculations. Nor is it hard to see how such a science would become embroidered with myth and fancy: a series of coincidences could rapidly assume the status of a natural law, and if religion were mixed up in it (which it usually was, because most early astronomers were also priests), then the actual scientific basis could be almost completely obscured.

Many primitive societies lived in a state of perpetual fear, because life was always marginal and the slightest deviation from a 'good' year could mean famine or death. Partly in order to please the gods whom they imagined controlled the planets, and partly in order to bolster their own power, the priest-astronomers often used to insist on sacrifices. At first, they used to sacrifice human beings: a beautiful virgin was thought to be most pleasing to the gods, because she was most pleasing to man, but then the sacrifice declined to animals, and now in our own time it is symbolised by the wafer and the wine at a Christian communion, or the mandala and water offerings at a Buddhist shrine. The underlying point of all of these 'sacrifices' is to keep people aware of their religious duties, and of the essential relationship of man to the cosmos.

The regular, ordered world of the skies was a reflection of the regular, ordered world in which our ancestors wanted to live; and a comet, or 'hairy star', which came from outside the Heavenly Spheres was a symbol of the disruption of this ordered world. Small wonder, then, that it was seen as an omen of doom, and that all sorts of magical rites were performed to turn aside disaster. And, in most societies, it was the astrologers who were relied upon to say *when* (and sometimes *how*) these rites were to be performed.

Most people do not realise it, but astrology has two faces: the intuitive, mystical face, which is what we see when we read our horoscopes, and a surprisingly rigorous mathematical face, in which the angles of the planets relative to one another, to the Earth and to the sun are all carefully calculated. This is one of the reasons why serious astrologers (by which I mean astrologers who take their work seriously, whether others do or not) just laugh at the daily 'horoscopes' which appear in newspapers: at best, these can only be wild generalisations, and at worst (which is what they usually are), they are utterly useless. Take a look at 'your' horoscope for today; can you seriously believe that one-twelfth of all mankind, spread across the whole globe, will all have 'a financial problem solved', or will find it 'a good time for travel'? Of course not; but equally, if you have ever had an individual horoscope cast, have you not been surprised — sometimes uncomfortably — at its accuracy?

The astrologers, though, never managed to fit comets into the mathematical side of their art — and indeed, they were never much concerned with formulating a general theory of planetary motion, because they were concerned with practical prediction, not with inapplicable abstractions, which is hard for us to appreciate from a twentieth-century standpoint. It was therefore left to astronomers to devise the heliocentric (sun-centred) model of the solar system, and to discover mathematical laws which could be made to describe the motion of the planets.

Astronomers were a part of the new phenomenon which arose at the time of the Renaissance. Until then, the classical authorities such as Aristotle and the Church Fathers had reigned unchallenged in the West for one and a half millennia; their word was law, and it was considered mere foolishness to dispute it. With the Renaissance, though, there came scientists who sought to discover truth for its own sake, unprejudiced by old ideas and beliefs, even though this often brought them into conflict with the Church. The turning-point in astronomy was the publication in 1543 of Galileo Galilei's *De Revolutionibus Orbium Colelestium (On the Revolutions of the Heavenly Spheres),* which stated that the Earth was *not* the centre of the universe, but moved in a circular orbit around the sun — an argument which was in direct contradiction to the teachings of the Christian Church, and which led for Galileo to an extremely unpleasant run-in with the Inquisition, including threats of torture, and an enforced public recantation of his beliefs.

These men of a new discipline tried to find things out for themselves, rather than going back to classical texts; this is what we now call the 'scientific method'. Most tested their ideas by experiment or observation; anyone who was caught out in something which was demonstrably wrong could no longer take refuge in the writings of classical authors, or in the teachings of the church, but had instead to defend his point of view using only logic and observable facts. The Dane Tycho (or Tyge) Brahe (1546–1601) was one of the great astronomical observers. He was a colourful character, who wore a gold nose to replace the original (lost in a duel) and wrote such appalling poetry that he is reputed to have jailed (in turn): the publishers who refused to publish it; the printers who refused to print it when he wanted to publish it himself; and finally, the paper-makers who refused to supply him with paper! He was, however, a superb observer, and his observations were used by the German astronomer Johannes Kepler (1571–1630), who modified Galileo's model by substituting ellipses — slightly flattened circles — for the circles which the Italian had suggested. Kepler himself was not at all happy about this, because the Church had accepted the Aristotelian idea of the 'perfect' circle, and the idea that God would use an 'imperfect' shape like an ellipse was vaguely blasphemous; it implied that the Supreme Being was less than perfect — or even that He was a little bit sloppy.

Kepler found, though, that the model of an ellipse worked perfectly for all of the planets then known (it also works for the ones which have been discovered since), so he wrote his First Law of Planetary Motion: *Planets revolve in elliptic orbits about the sun, which occupies the common focus of all these orbits.*

His Second Law is equally important. If the orbits are elliptical, then the speed (to be accurate, orbital velocity) of the planets is not constant; it will depend on where on the ellipse they are. This is formulated in the law which states: *If a line is drawn from the centre of the sun to any planet, this line as it is carried forward by the planet sweeps over equal areas in equal portions of time.* These two Laws appeared in Kepler's *Astronomia Nova Aitiologetos* in 1609.

Although the English knight Sir William Lower realised as early as 1610 that these two laws could also be used to solve 'the unknowne walks of comets', Kepler himself believed that comets always travelled in straight lines, even though he continued to work on astronomical theory and evolved a third law (published in 1619) which said: *The squares of the periods in which the planets describe their orbits are proportional to the cubes of their mean distances from the sun.*

Geniuses though many of these men were, their work was to be the foundation for the work of another scientist who came after them, a man whom many reckon to be the greatest scientist of all time. He was Sir Isaac Newton (1643—1727), whose *De Motu Corporum in Gyrum* (*On the Motion of Bodies in Circles*, 1684) provided mathematical demonstrations or proofs of Kepler's three laws, and laid the foundation of his own monumental *Principia Mathematica* which appeared just over two years later. Newton's contribution to classical science was so great that to this day scientists talk of 'Newtonian' physics, though many of his models have been superseded by newer physics based on the work of Einstein, Planck, Bohr, Heisenberg, Pauli and others.

It is important at this point to understand the nature of science. Nowadays, only a very bad (or very old-fashioned) scientist would pretend that Science (with a capital 'S') Had All The Answers. Science is not reality itself; it is a description of reality, a description which works better than any other that is available. Some writers attempt to discredit Newton, by saying that he is 'wrong' in the light of modern physics; most physicists, on the other hand, will use a Newtonian model if that works best, and another model if that fits the facts better than Newton's theories. That one word, *theories*, is the heart of science: the moment you lose sight of the fact that all science is theory, and that the moon (or at least those parts of it which have not been explored) may be made of cheese, you cease to be a scientist. It is perfectly all right to assign relative values to theories, and to rank the theory that the moon is made of cheese a long way below the theory that it is made of rock; you can even discard the cheese theory completely, because you consider it so improbable as not to be worth considering. But you must be equally ready to discard your more precious and carefully-reasoned theories if some other theory arises which fits the facts better, and many so-called 'scientists' are not willing to do that.

Newtonian physics is not a subject which can be covered in a few lines, or even in a few books, but for our purposes the most important thing is the old chestnut about the apple (if such a vegetarian cocktail of metaphors makes sense). It was obvious that there had to be some form of attractive force, which made apples fall to the ground, and many other scientists had proposed theories to explain it. Newton's contribution was that he proposed a mathematical model (which works) to explain the way in which apples fall to the ground, whereas previous scientists and philosophers had either sidestepped the question by saying that 'things fall because it is their nature to fall' (which is akin to saying that we live because we live) or had devised the most wildly complicated and mathematically insupportable ideas, such as Descartes' theory of 'vortices', a complex tangle of gravitational whirlpools.

Newton's theory that gravity obeyed an *inverse square* law has worked very satisfactorily since he first suggested it: an inverse square law is simply one in which a force weakens according to the square of its distance from its source (or, in the case of an interaction between two bodies, according to the square of their distance apart). Light, for example, obeys an inverse square law: if there are three objects, placed one foot, two feet and three feet respectively from a lamp, then the one at two feet will receive one quarter as much light (because four is two squared) as the one at one foot, and the one at three feet will receive one ninth as much light (three squared is nine). Admittedly, the idea that gravity might be proportionate to distance was not a new one — both Hooke and Halley had suggested as much — and Halley had suggested an inverse square law, but it was Newton's mathematical ability which demonstrated such a relationship, as well as promoting the idea (which was revolutionary at the time) that gravity might be a universal force, the force which held the planets in their orbits around the sun and the moon around the earth, as well as a terrestrial phenomenon.

This inverse square law of gravitation explained Kepler's laws of planetary motion beautifully, and when combined with Newton's own Three Laws of Motion (all the best laws seem to come in threes, except Jewish commandments), it laid the basis of Newtonian physics. Newton's Three Laws of Motion are, of course:

1 A body remains in a uniform state of motion unless it is acted upon by an external force.
2 The force applied to a body is in the direction of the acceleration imparted to the body, and is equal to the mass of the body times its acceleration.
3 Every action has an equal and opposite reaction.

Now — at last — we get to Edmond Halley, who may have pronounced his name Halley (as in 'Alley'), Hayley (as in Mills or Bill), or Hawley: attempts to find out which are pretty much false scholarship, the more so as he spelled his name in several different ways, as was the custom before dictionaries were invented and spellings consequently standardised.

Born in 1656, Halley was the son of a wealthy London soap-boiler, and he was determined to make his mark as quickly as possible upon London's scientific establishment of the time: science was very fashionable, and Charles II lent his patronage to Britain's leading scientific institution, the Royal Society. Halley was apparently dynamic to the point of pushiness, but well-liked; after St Paul's School (where he was School Captain, as well as a skilled experimenter and constructor of 'caelestiall Globes' or star maps), he went up to Queen's College Oxford in 1673, when he was seventeen, and in 1676 he set out to map the stars of the Southern Hemisphere, something which no-one had done at all thoroughly to date, and which he seems to have undertaken as a calculated exercise in building his own reputation — which was already pretty good for a twenty-year-old, as he had become a protegé of Flamsteed, the Astronomer Royal, while still at university, and was known as a formidable mathematician. By the age of twenty-two, he had gained his Oxford MA, as well as his ambition of election to the Royal Society.

But where could he go from there? The thing that made his name a household word, instead of condemning him to the ranks of scientific giants such as Hooke who were overshadowed by Newton, was his interest in comets. It was first aroused by a comet which appeared in November 1680. As is usual with comets, it was visible for a few weeks (or months if you had a telescope), then invisible for a few days at perihelion (the nearest approach to the sun), and then visible again afterwards; as it approached and departed, it was apparently a magnificent sight with its tail stretching across 70° of the night sky. The thing was, though, that most people took this dual sighting to be the sighting of *two different comets*.

By coincidence, there *had* been two separate comets in late 1664 and early 1665, which is probably why they thought this, but Flamsteed and young Halley suspected that they were seeing two manifestations of the same thing. Meanwhile, up at Cambridge, Newton was also observing the comet: he suspected that Kepler's theories about straight-line motion were wrong, but he needed observational data to be sure.

The scientific community of the period was close-knit, even if it was periodically riven with quarrels and personality clashes, so it was entirely natural that Halley and Newton should meet sooner or later; and it was in August 1684 that

Halley made the journey up to Cambridge to see Newton, in order to ask for help in a mathematical proof of his pet inverse square law theory. Hooke (of Hooke's Law fame) claimed to have worked out such a proof, but was unwilling to produce it when challenged, so Wren (the same Wren who designed St Paul's, and who was a close friend of Halley's) called his bluff by offering a small prize — a book — to the first of the two to produce an actual proof. Hooke never succeeded, and Halley was unable to do it alone, so he asked Newton for help.

We have already seen that Newton did in fact produce such a proof, later in the same year. It is likely that he had already worked it out, but that he was afraid of Hooke's getting in on the act, because the two men disliked one another intensely. Once he was satisfied that Halley had not been sent by Hooke, he gave him the proof. Halley was so impressed by the mathematical genius of the man who could produce such a proof that it was he who urged Newton to write the book which he knew he had in him — the book in question, of course, being the *Principia Mathematica*. For this alone, regardless of Halley's own genius, his name deserves to live forever.

But to return to the comet, Newton and Halley collaborated to a great extent on investigation of the laws of cometary motion. Newton was within an ace of solving the problem on his own, using his own observations, but he was handicapped by a lack of observational data: the comet of 1680, sometimes called 'Newton's Comet' was almost all that he had to go on. Although he managed to shake off Kepler's idea of straight-line motion, and postulated the theory that comets were subject to exactly the same laws of planetary motion as the planets (albeit with very much more eccentric orbits), he stuck to the idea of a parabolic orbit rather than an elliptical one; a parabola is one end of an ellipse of infinite length. In any case, the 1680 comet simply did not provide enough experimental data, and Newton's figures were not always exact.

Halley was both luckier and more thorough. His erstwhile friend and mentor Flamsteed, who was apparently a short-tempered man and with whom he had by now fallen out, had made meticulous observations of another comet in 1682, and Halley managed (by somewhat devious means) to get hold of Flamsteed's figures. He also researched the history of comets as far as he could, looking for evidence of periodical return; we know that he picked up this idea of periodicity in Paris in late 1680 or early 1681, from the Director of the Paris Observatory, Giovanni Cassini, though Cassini believed in a very much shorter period than most comets demonstrate — a mere two or three years. It is also important to remember that Halley had almost two hundred years of fairly reliable astronomical data to rely upon, though obviously the reliability was variable and tended to diminish as he reached further and further back, but the course of the brightest comets had been charted through the constellations and provided a fair indication of the individual orbits — which would, of course, be constant if they were of the same comet.

Two and a half centuries later, Einstein was to say that chance favours the prepared mind — and Halley was both lucky and well prepared. The orbit of the 1682 comet was remarkably similar to that of a comet observed by Kepler in 1607, and more than adequately similar to one seen by Peter Apian in 1531; he also had reports of a bright comet in 1456, and, although the standards of observation were far lower then, it looked like the same comet, with a period of 75 or 76 years. After endless calculation, a very tiresome business in the days before computers or even slide-rules, he finally published *Astronomie Cometicae Synopsis* in 1705, with a parallel English edition entitled *A Synopsis of the Astronomy of Comets*. In it, he predicted the return of the comet which now bears his name. The stage was set for modern cometography — and also for astrologers who could now compute the possibility of disasters well in advance, without having to wait for the surprise appearance of the 'hairy stars'.

3
THE STAR-CROSSED SKY

When Halley published his *Synopsis* in 1705, he was forty-nine years old, and 'his' comet would not appear again until 1757 or 1758; he would have to live to be over a hundred to see his theory vindicated. In fact, he died at the age of eighty-five in 1742, after a long and adventurous life which had included not only giving his name to the first comet whose return was accurately predicted, becoming the Savilian Professor of Geometry at Oxford in 1703, and succeeding Flamsteed as the Astronomer Royal in 1720, but also such diversions as being the Deputy Comptroller of the Chester Mint, some Secret Service work for the Crown, hobnobbing (and drinking a great deal of brandy) with Tsar Peter the Great on his visit to London in 1698, and commanding a Royal Navy expedition into the South Seas; until he received his honorary doctorate in 1710, he was normally referred to as 'Captain Halley', and apparently could swear like a sea-captain to prove his claim to the title.

In due course he was vindicated, though, by a French astronomer called Clairaut, a clockmaker's wife called Nicole Lepaute and the Royal Astronomer to the French Court, Joseph de Lalande. Clairaut was the moving force, and he and Lalande were assisted in their calculations by Mme Lepaute, who was apparently a formidable *calculatrice* and without whose help the two men admitted they could never have done the work. Halley had predicted that the extremely eccentric orbit of any comet would mean that its period could be delayed or accelerated by passing near the planets of the solar system, especially the giants Saturn and Jupiter: if it passed them on the way 'out', its return would be accelerated, as its outward passage would be shortened; if it passed on the way 'in', it would be slowed. Although Halley had predicted this, he had not applied himself to the actual mathematics: the mammoth task of calculating the precise date of perihelion actually took the French trio until November 1758, by which time the comet had still not appeared. People were beginning to scoff, especially the French who were sure that if anyone were to predict the return of a comet, it would be a Frenchman, but the calculations indicated a delay of over 600 days, or nearly two years, so that the actual perihelion would be in 1759. The comet was first sighted on the evening of Christmas Day 1758, by George Palitsch of Prohlis, near Dresden, and perihelion was on exactly the date which the French team had predicted using Halley's figures and the best observational data available, 12th March.

The 1835-6 return also appeared on schedule, and so did the 1910 return, but that is dealt with in the next chapter. Now that we know that Halley was right about comets, it is time to look a little more closely at what he didn't know — and what we ourselves still do not know for sure. The rest of this chapter is accordingly concerned with three things: first, with what a comet is; second, with the number of comets which pass close enough to the earth to identify in any one year; and third, with the possible mechanisms by which comets could influence life on Earth.

The answer to the question of what a comet is is simple: nobody knows. The two most popular models, though, are the *dirty snowball* and the *flying sandbank*.

The 'dirty snowball' model describes the core or nucleus of the comet as a central mass of uncertain composition (but probably mostly light elements) surrounded by a dusty frozen cocktail of variously unsavoury compounds, principally water (H_2O), ammonia (NH_3), and methane (CH_2), but also including other simple and not-so-simple chemicals such as carbon dioxide (CO_2), dicyanogen (C_2H_2), and possibly even such complex molecules as ethyl alcohol (C_2H_5OH). The idea that the nucleus might be a frozen conglomerate dates back to the last century, but its present form is due to the American astronomer Fred Whipple. The 'snowball' itself would probably be rotating, and its temperature in the outer reaches of the solar system (or even beyond it) would be close to absolute zero, about −273°C. Only as it approached the sun would some of the snowball tend to boil off (actually 'sublimating' straight from the solid to the vapour state, without ever passing through the liquid phase), and it is this boiled-off material which constitutes the tail. The 'flying sandbank' model sees the nucleus of the comet as a collection of separate particles, travelling together as a loose-knit whole. Although it was popular in the nineteenth century, this view no is longer widely held, though it is still defended by some very eminent atronomers.

Either model allows for the break-up of comets, which

has been observed on several occasions, and there is no doubt that comets 'wear out' as they repeatedly pass the sun, each time losing a little more of their mass. It may well be that some meteors are cometary debris, and we shall return to this in Chapter 9; the likelihood is that if there were ever a cometary collision with the Earth, it would be a *part* of a comet, rather than a whole comet.

Whatever the construction of the nucleus, the make-up of the envelope is rather more certain, because we can check it by astro-spectrophotography; the word 'envelope' is used instead of 'tail', because although the tail is the most visible part there is actually an 'atmosphere' of the same material which extends either side of the nucleus and a little in front of it. The further from the comet you are, the thinner the envelope, but the material is essentially the same as in the tail, and this is material sublimated from the surface of the comet by the sun's heat. It is this which gives us our best idea of what the nucleus itself might consist of.

There are actually two types of tail, the gas tail and the dust tail. The gas tail is composed of gases which are in a state known as *plasma*, meaning that the solar wind (to which we shall return later) keeps their molecules in a state of constant agitation, rather like the neon in a neon light, and causes them to emit light. Gas tails are long and straight, and point directly away from the sun. The dust tail is made up of incredibly finely dispersed dust, and is usually scimitar-shaped, because it lags behind the comet's movement; the gas tail is actually slightly curved too, but much less detectably. It is important to realise just how thinly matter is spread in a comet's tail: on Earth, it would be regarded as a very 'hard' vacuum, and some idea of its lack of density can be gauged by imagining a cubic mile of space with perhaps thirty or forty small glass marbles spread all through the space, although of course the material would be very much more evenly spread than this, as a 'dust' so fine that it was composed not only of the minutest particles, but also of free molecules and even individual free atoms.

The answer to the question of the frequency of comets depends on how hard you look for them. There is no doubt that they are vastly more common than was once thought, and that at any one time there are likely to be ten or a dozen detectable, but the majority of them are so faint that they can only be seen through large and powerful telescopes. Amateur astronomers, with less powerful telescopes, can still see a dozen or more a year if they care to look, but comets which are bright enough to be seen with the naked eye occur much more infrequently. As we have already seen, we can rely on Halley's Comet every 74-76 years, but the last major comet before the 1985/86 appearance of Halley's Comet was Comet West in 1976; it far outshone Kohoutek, which was discovered by a Czech astronomer in March 1973 and predicted as a possible bright object for around Christmas 1974, but, because Kohoutek had been such a disappointment, most people did not bother to look for it, the more so as this would have involved getting up very early in the morning.

Before Kohoutek, there was Comet Bennett (1970) and Comet Tago-Sato-Kosaka (1969), both of which were more visible than Kohoutek. Tago-Sato-Kosaka gained its unwieldy name because it was spotted more-or-less simultaneously by three Japanese astronomers (the maximum number of names which can be tacked on to a comet is three), but it was a 'sungrazer' which approached so close to the sun that it actually skimmed the solar atmosphere — and broke into two pieces as it did so. In other words, there were four major comets in the sixteen years preceding Halley's

Comet, and (for example) we have already seen that there were bright comets in 1664, 1665, 1680, and 1682 — four comets in eighteen years — while Halley's last appearance in 1910 was followed by Delavan's comet in 1914. On the other hand, there are often quite long periods during which there are no bright comets visible at all: for about forty years, from Delavan's comet to the mid-1950s, there were none bright enough to capture the public imagination, despite the fact that there was no shortage of events for them to herald, including the deaths or abdications of three British kings, Hitler's rise to power, the Second World War, and the explosion of the first atom bombs.

Although the periods of comets are predictable, and new ones are appearing all the time, it is obvious that if the periods of comets range (as they do) from a few years to many thousands or even millions of years, not to mention the comets which apparently leave the solar system for ever, there will be times when they come in batches and times when they do not come at all — rather like buses. To take a simple example, imagine three comets with periods of 60, 100, and 150 years. If they all appeared together in one year, comet A would appear 60 years later; comet B would appear 100 years later; comet A again 120 years later; comet C 150 years later; comet A for the third time 180 years later; and so on until A and B appeared together 300 years later and all three together again 600 years later. In practice, the periods are neither as regular nor as simply related as this, so wide variations from decade to decade (or even century to century) in cometary frequency are quite normal.

Now we come to the nub of this book: how could comets influence events here on Earth?

Perhaps the most important theory is the one which ties together all known methods of prediction, namely Jung's theory of *synchronicity*. Carl Gustav Jung (1875-1961) was a Swiss psychologist, psychiatrist and writer who formulated his theory after being struck by the extraordinary nature of coincidence. Without going into Jung's theory in depth, it is a fair summary to say that he believed in the seamlessness and interdependence of the whole universe; the 'scientific method', which seeks to isolate chunks of reality and 'explain' them in terms of cause and effect is, according to this theory, inherently limited. Jung was very interested in the *I Ching*, an incredibly ancient Chinese method of divination using either yarrow stalks or coins, and he believed that one of the reasons why the I Ching 'works' for so many people is that, although selecting the yarrow stalks or throwing the coins may appear to rely on 'blind chance', there is no such thing as truly 'blind' chance: everything which happens in the cosmos is related to everything else, though not necessarily in ways that we can understand.

Unlike most theories, and certainly unlike any theory based on classical scientific method, the theory of synchronicity is *acausal*; that is, two (or more) events may be related, but they are not related in the familar pattern of 'cause-and-effect'. Instead, they are more like a coincidence, but so closely related that it is hard to imagine that they are 'pure' coincidence, with no link whatsoever.

Although acausality may seem to be mystical and self-defeating, it is actually quite close to several fundamental views of modern physics, with which most people are now familiar. To take one of the best known examples, subatomic 'particles' are now seen as having a dual nature, behaving sometimes as particles and sometimes as waves; hence the term 'wavicles' used by some writers. How they do this we don't know, and just to add to the fun, we do know that we cannot even describe the shape or location of a sub-atomic particle in terms which we can visualise.

The old concept, as taught in schools, of an atom as a miniature planetary system is a gross oversimplification; the electron orbits are seen as 'probability areas', where there is a strong probability of finding *something*, but we don't know what it is and we can't guarantee that it is there anyway. In any case, when we refer to a 'wave', we are using an analogy, because we don't know what it is a 'wave' in. On top of all this, there are several phenomena which we can *describe* mathematically, but which we cannot *explain*; gravity is an excellent example of this. We know that when there are two bodies there is always an attracion between them — but we don't know how it works (although we can describe and predict its behaviour), so we could call that an 'acausal' phenomenon as well!

Nevertheless, we also know that, although we can be certain of virtually nothing when referring to a single subatomic particle, we can talk with considerable certainty about what will happen when we consider large numbers of atoms — which, of course, is what actually happens in everyday life. When we consider (say) a chemical or nuclear reaction, we know that we are dealing with atoms in sufficient numbers that our statistical predictions will be accurate. And, to reiterate, we can describe and predict the effects of gravity, even though we cannot explain how gravity works.

It is a crude and debatable analogy, but we can say the same to a certain extent of astrology: although we do not know *why* or *how* the stars (or to be more accurate, the planets, the moon, and our personal star, the sun) might influence our lives, there is at least a case to be made that they do. Even if you discard astrology completely, the influence of the moon cannot be denied: ask anyone who works in a mental hospital. It is true that the gravitational and (probably) the electromagnetic influence of a comet is far less than that of the moon, but it is a moot point whether it can actually be ignored.

This brings us to the next possibility of how a comet could influence the Earth. Instead of dealing with mystical and acausal considerations of the unity of the universe, we deal with three well-known and scientifically measurable phenomena: gravity, electromagnetics, and the solar wind.

We have already said that for all Newton's physics, or Einstein's for that matter, we do not have any clear idea of *what gravity is:* we do not have anything equivalent to the atomic theory of matter or the wave theory of electromagnetic radiation (or the 'wavicles' of modern science) which explains *why* it works. Even where we can make accurate predictions, there are gaps in our knowledge: for example, the moon apparently has local gravitational anomalies, which are usually explained as the result of mass concentrations ('mascons') under the surface — but we don't know what these 'mascons' are. Even on Earth, there are places where gravity does not seem to behave itself properly, though most of these are the result of optical illusions, where the slope of the hills makes an up-gradient look like a down-gradient, so cars apparently 'roll uphill' when the brake is released. Others are less easily explained, though they have come in for surprisingly little scientific investigation.

One theory about the way in which birds navigate on their seasonal migrations is that they have some sort of 'gravity sensors', which allow them to orient themselves with respect to the Earth's (and possibly the moon's) gravitational fields. If this is so (and it is only a theory — remember what was said about theories in the last chapter), then it is quite possible that we are equipped with something analogous. The gravitic fluctuation caused by a comet, even a comet passing much closer to the Earth than Halley's Comet has ever done, is minute, but we do not know *how* minute a variation we can detect — if we can detect one at all.

We do however know that we can detect electromagnetic waves over a vast range of frequencies, from the long waves which we call radio waves, through the narrow spectrum which we call light, through to the X-rays which are harder and harder to stop as their wavelength grows shorter. We know that the Earth has an electromagnetic field, which we very conveniently use for navigation, and most of us remember from our school days that there is a difference between 'true North' and 'magnetic North', generally known as 'compass deviation'. What is not widely known, at least to the general public, is that the rate of drift of the magnetic north from the true north has increased dramatically since about 1950, and that navigational charts are now re-issued every five years in order to allow for this. From 1850 to 1950, the average annual drift of the magnetic north in relation to the true north was about two miles a year; now, it is more like *ten times* that. Why? We don't know. What effect will it have? We don't know. It may well be that this increased rate of drift simply represents a return to a normal figure, after a century or more of unusual stability; it may also be that the polarisation of the planet is about to reverse, as it has demonstrably done in the past. What would that mean? Again, we don't know. Was it reversals of the Earth's magnetic field that put an end to several lines of evolution (including the dinosaurs), or was it a comet collision (or a series of comet collisions), as suggested in Chapter 9? We don't know.

The possibility that birds navigate by gravitation has already been canvassed, and the possibility that they navigate by reference to the Earth's magnetic field is another theory;

bees almost certainly use magnetic particles in their bodies to detect variations in the Earth's magnetic field, and use this for timekeeping as well as for navigation. However, other animals may sense electromagnetism; we know that we humans can do it, as is witnessed by the reactions which some people (not all) have to walking under high-tension power cables; in many countries, it is illegal to build power cables over existing houses, partly on safety grounds (what would happen if a cable snapped?) and partly because of these adverse effects, which can include headaches, depression, and nervousness. How do we sense electromagnetism, and how much does it take to affect our behaviour? Again, we don't know.

Nor do we know a great deal about the electromagnetic influence of comets on the Earth's magnetic field, except that it is not great. Most would say that it is negligible: they may be right. But then again, they may not be . . . .

The third effect, the so-called 'solar wind', is one of which many people will not have heard. Essentially, it is the endless stream of charged particles which issue forth from the sun all the time. The pressure of this 'wind' is tiny, but not negligible: it would actually be possible, in interplanetary space, to build a 'solar yacht' with sails of plastic thinner than gossamer, square miles in area, and 'sail' before the solar wind — but it would be an outward journey only, as there would be no way of tacking back against the wind. The solar wind is also strong enough to propel tiny particles — even particles as big as bacteria — through interplanetary or interstellar space, and this is something to which we shall return in Chapter 8.

It is the solar wind which is responsible for the plasma tail of a comet. The interaction between the solar wind and the material of the 'dirty snowball' (which is being boiled off by the sun as the comet approaches perihelion) gives rise to the plasma, which the solar wind then blows away from the comet, like the plume of steam behind a train or an aircraft's jet trail. The plasma tail points almost directly away from the sun, though there is a slight deviation caused by the speed of the comet.

The solar wind is responsible for such atmospheric effects on Earth as the Aurora Borealis and Aurora Australis, and for the so called 'sun storms' or 'radio storms' which periodically disrupt radio communications. If the Earth passes through the plasma tail of a comet, these atmospheric effects could be expected. Here, although there is little likelihood of *direct* effects on human beings, the possibilities for *indirect* effects are tremendous: the ultimate possibility for a failure of radio-communications could even be a nuclear war, if signals from 'the other side' were misinterpreted by either the Eastern or Western blocs — or if, worse still, they were used as a blanket to conceal the launch of a nuclear attack.... As a matter of interest, when a defector brought a Russian fighter plane over to the West, observers first dismissed the apparently crude electronics (based on miniaturised valves) as yet another example of Soviet military backwardness. Then they realised that in the event of a nuclear war, the more sophisticated transistors would be knocked out at least temporarily, and possibly permanently. There is now a considerably revived interest in traditional valve-type electronics in the West.

Passing through the dust tail, on the other hand, could have completely different effects. As already mentioned, the dust tail is very tenuous indeed, and the risk of physical damage is zero. It does, however, bring us to the third way in which a comet could affect life here on what Hendrix called 'the third stone from the sun'.

The 'flying sandbank' and 'dirty snowball' models of cometary construction are not the only ones. There is a third, hardly accepted at all in the scientific community, but not actually refutable, which says that comets can actually carry some form of life — something viral in nature, perhaps, which could infect the Earth and everything in it. According to this theory, many diseases are literally extraterrestrial in origin; and life on Earth itself may have originated from these beginnings.

The theory was first postulated by two astronomers, Sir Fred Hoyle and Professor Chandra Wickramasinghe, and, although it is easy to dismiss it as 'unscientific', it is actually a lot less 'scientific' to ignore a theory simply because you do not like it rather than to keep an open mind and to admit the possibility — however minute — that it might have some merit. We shall come back to this theory rather more carefully in Chapter 8.

Our final theory for comets affecting the earth is simply the *self-fulfilling prophecy*. In other words, if you think that something terrible is going to happen, it may well do so. Self-fulfilling prophecies work in three ways. The first, and probably the least important, is *obsession*. This is simply a form of mental illness in which the sufferer thinks about one thing to the exclusion of all else; in its extreme examples, it is seen in the man who jumps from a skyscraper window so that the comet won't 'get' him (which, of course, it has), or in Nero, who massacred anyone who might present a threat to his throne because he saw a comet which might have presaged his own death . . . .

After obsession comes *distraction*, which is probably a much more potent force. If we feel generally uneasy, our concentration is likely to be reduced, and if our concentration is reduced, we are more likely to have accidents, especially if we are driving cars, trains or aircraft at the time. Of course, a book like this can be accused of contributing to the effects of distraction — which is why I hope you will reach Chapter 10, where we look at ways of counteracting 'comet fever'. And after obsession and distraction comes *coincidence*, which actually takes two forms in its own right. One is 'pure' coincidence, where two events occurring simultaneously or one after the other seem so inseparable as to be undeniably linked, and the other is self-delusional coincidence, where the link is not one which people would normally automatically make, but which is prompted by distraction; indeed, distraction and coincidence can feed upon one another until they amount to obsession, which is as neat an example of three-in-one and one-in-three as you could wish to see outside the Holy Trinity.

In practice, any or all of these theories could operate independently or simultaneously; the link between synchronicity and the self-fulfilling prophecy is so close as to be virtually circular, and the others can be tied in equally neatly. In Chapters 5-8, we shall look at the ways in which Halley's Comet may have influenced world events in the eighteen months or so preceding perihelion, and in the final chapter we shall look at what more Halley's Comet may have in store for us. But, before we go on to that, it is appropriate that there should be one more historical chapter: a brief look at the events which have attended previous apparitions of Halley's Comet.

4
PAST APPARITIONS

The astronomer's word for the appearance of anything in the sky is 'apparition' — a word which simply means 'appearance', but which is curiously appropriate when applied to comets, because it is also used in English to describe a ghost or other psychic manifestation.

Various astronomers have calculated the exact dates of perihelion of Halley's Comet for the last 2,000 years, and although it has been traced back to 240 BC, the first important historical date to which attempts have been made to link it is the birth of Christ — in other words, to make Halley's Comet the Star of Bethlehem. Unfortunately, the dates just do not mesh: the nearest possible apparition would have been in 12 BC, and although modern scholarship (rather confusingly) puts Christ's birth in about 7 BC, a discrepancy of about five years does seem excessive. A traditional link with kings does however appear: in 12 BC, Herod the Great was granted the title King of Judea by the Romans — the same Herod who ordered the Massacre of the Innocents. A much more credible explanation of the Star of Bethlehem was that it was a conjunction of the planets, namely Jupiter and Saturn in the constellation of Pisces, which would (to a trained astrologer, especially one looking for a Messiah) clearly point to a Jewish leader's being born in Palestine. Giotto's oft-quoted *Adoration of the Magi* does show Halley's Comet, but only because he saw the 1301 apparition.

The next five apparitions, in AD 66, 141, 218, 295, and 374 can be tied in with the general decline of the Roman Empire, especially if you accept Hoyle and Wickramasinghe's disease-bearing theory.

Rome's burning in July 64 was within eighteen months of the 66 apparition; a revolt in Judea in 66 wiped out a Roman garrison, and the Britons under Boadicea (Boudicca) were rebelling too; and Saints Peter and Paul were both executed around this time. The 218 apparition coincides with Heliogabalus' (Elagabalus) elevation to Emperor of Rome: one of the most nauseating of all the decadent emperors, he was a teenage boy who was given licence to live out all his adolescent fantasies (such as having his carriage pulled by naked women, or of literally smothering his mistresses in rose-petals), and his rule was short (218-222) but horrific. On the other side of the world, it was a time of civil war in China, but this is hardly unusual.

By 295 the Roman empire was staggering on its last legs, divided into the Eastern and Western Empires, with two Emperors. A kingly link appears in the accession of Narses to the Persian throne, and the 374 apparition corresponds to the time when Gratian became Emperor of the West upon the death of Valentinian; the Goths overran the Empire of the East and killed the Emperor Valens at the Battle of Adrianopolis.

From the 374 apparition to the 989 apparition, we are in the formlessness of the Dark Ages, at least in the West, but it is, worth noting that the Battle of Chalons (where Attila the Hun was defeated by Aetius, and anything up to 150,000 men were killed) coincided with the 451 apparition and that the advance of the Goths which preceded their sack of Rome in 537 must have begun at around the time of the 530 apparition. The harsh winter of 760/61 was blamed on Halley's Comet, which appeared in that year, and, on the regal front, Offa became King of Mercia and King Pepin the Short of the Franks died and was succeeded by Charlemagne and Carolman. Louis VI of France died shortly after the 837 sighting, and a little later Ethelwuld succeeded Egbert as King of Wessex. For the 912 sighting, a trio of monarchs died: Halfdan, King of York, Ethelred, King of Mercia, and Leo VI, Emperor of Byzantium.

In China, there were reliable astronomers throughout the period, but their job was generally separate from that of the astrologers, who processed the astronomers' data, and although we have the raw figures we do not have much in the way of prediction, confirmed or otherwise. As with Roman data, there is probably a rich source of material here, waiting to be explored, though it may well have been destroyed during the Cultural Revolution; the same would be true of Tibetan sources, which were reputed to stretch back as far as the Zhangzhung kings of two and a half millennia ago, but which were almost certainly destroyed by the Chinese after the 1949 invasion. Again, there would be a tremendous amount of Arab, Greek, and possibly earlier information, if the great libraries of the Near East had not been destroyed by both Moslem and Christian fanatics and bigots between 700 and 1000 years ago. To any historian, the loss of knowledge occasioned by political and religious bigotry is heart-breaking; Hitler's book-burning was mere amateurism compared with the loss of the libraries of Alexandria and Constantinople, many of the books in which were unique manuscripts of great age.

In 1066, though, comes an event which any student of English history will recall — the Battle of Hastings — and the appearance of Halley's Comet is actually recorded on the Bayeux Tapestry, the immortal record of the invasion commissioned by William the Conqueror to commemorate his success.

The 1145 apparition coincided with the Second Crusade, and 1222 ties in well with Genghis Khan's sweep through Afghanistan, India, and China. Intriguingly, Genghis Khan regarded comets as his special stars, and is said to have embraced Tibetan Buddhism; although he is held up as a butcher and a barbarian by most histories, it is worth reflecting that most of our accounts come from the Persians and the Chinese, both of whom were openly scornful of the great Khan until he thrashed them conclusively, and neither of whom has ever been at a loss for words when it comes to massacres or atrocities on their own account, or false propaganda about their opponents either for that matter.

The 1301 apparition marked a period of general imbroglio world wide, and the 1378 apparition marked the death of England's Edward III and his replacement with Richard II, who almost immediately had to face the Peasant's Revolt under Wat Tyler; in the same year Pope Gregory IX died, and after his death the Great Schism (during which there was a second Pope at Avignon) began. Outside Europe, the 1378 apparition must also have appeared as the star of Timur the Lame, or Tamerlane (Tamburlaine), who by all accounts *was* as unpleasant as history has painted him. The 1456 apparition coincided with the beginning of the Wars of the Roses in England, and the 1531 apparition presaged the end of the Hanseatic League, which changed the shape of Northern European politics, and marked Henry VIII's break with Rome (which actually finally occurred in January 1535, but which had been brewing for some time). It was also the time of Pissarro's invasion of South America, and it unquestionably coincided with the great earthquake of Lisbon, in Portugal, when 30,000 died, and another near Naples in which 60,000 or more died.

To mark the 1607 apparition, Charles IX was crowned King of Sweden, marking one of the more warlike epochs in that country's history, and the Spanish revolt against the Moors got under way; it was to result in the expulsion of the Moors eleven months after perihelion. There were also riots in England over land enclosure.

The 1682 apparition had further regal connotations, with the death of Charles II and the succession of James II in England, and the recognition of Louis XIV, 'Le Roi Soleil', as the king of all France, but its most important effect was that it laid the foundations of Halley's work on comets, as we have already seen. The 1759 apparition coincided with with an earthquake in Syria, which killed more than 20,000 people, as well as with the fall of Quebec; a year later, George II of England died and was succeeded by George III.

The apparition of 1835 heralded a politically important period, in which Ferdinand I succeeded Frances I as Emperor of Austria and Peel resigned as the British Prime Minister, to be replaced by Lord Melbourne; a more forcible redistribution of political power was attempted by a man who tried to kill America's President Jackson, but he failed; and, at the time, the United States was at war with Mexico (the Alamo was in February 1836), and the Abolitionists were rioting in New York City, where a great fire later in the year (in December) caused over $18,000,000 worth of damage. Of course, Queen Victoria came to the throne shortly after this apparition, too. It was a good year for astronomers: both Simon Newcombe and Giovanni Schiaparelli were born in March. The Church of Jesus Christ of Latterday Saints, better known as the Mormons, was gathering momentum, too: formed by Joseph Smith, the first temple was incorporated in Kirtland, Ohio, in March 1836. This ties in with our observations elsewhere about the rise of charismatic churches in strange times. Finally, the eruption of Vesuvius in August 1834 was within our self-imposed 18-month time limit from the 1835 apparition.

The 1910 apparition coincided with the death of Edward VII and the succession of George V in England, the assassination of Prince Ito of Japan in late 1909, and a mining disaster in Illinois which killed 259 people. On this occasion the comet came much closer to the Earth than in 1986, and was therefore much more visible.

It is also interesting to note that, although the above information relates to Halley's Comet alone, sightings of other comets have marked events such as the Great Plague of London and the Great War of 1914—19.

But what of the period immediately preceding the 1986 visit of Halley's Comet?

5
EARTHQUAKES AND WEATHER

Unless you have been to Mexico City, it is almost impossible to describe. It is Chicago crossed with Old Delhi; it is a dream crossed with a nightmare; it is the past crossed with the future. Built on the soft soil of a dry lake bed, so soft that all major buildings have to be supported on huge piles, it combines first-class hotels and the Mexican headquarters of multinational corporations with slums, shacks, and squalor which are literally unimaginable to a Westerner unfamiliar with the Third World. You can get some idea of Mexico City from the pictures, but they cannot convey the smell of heat and dirt, the heartbreakingly beautiful eyes of the child beggars, or the way in which walking through the air-conditioned doors of a Western-style building is like walking from one country to another: the slamming heat, blinding glare, and frenetic rush of the outside is transformed into an oasis of coolness, shade, and tranquillity.

Mexico City is the centre of a sprawling urban conglomeration which covers about one per cent of the country of which it is capital, but which contains about eighteen million people, one quarter of the entire Mexican population. It is a roaring, bustling, hustling, dirty energetic city which suffers from near-terminal air pollution as a result of the Mexican government's cheap gasoline policy.

Since eighteen minutes past seven on the morning of Thursday, 26th September, it has been a ruin.

A disaster on the scale of the Mexico City earthquake cannot be conveyed *in toto*; instead, it can only be shown on a human scale. We have all seen the pathetic bundles — some dead, some alive, some dying — brought out of the wreckage of the maternity hospital, but to say that there are 10,000 or 15,000 or 20,000 dead, or that damage amounts to two billion pounds, means nothing against an image of a man scrabbling in a 100-foot-high pile of rubble which was once a thirteen-storey block of public-housing flats, which housed his wife and three children when it collapsed. It means nothing against a child dazedly cuddling a puppy which is all that remains of his family: mother, father, brothers and sisters are all lost. It means nothing against a woman slipping, sliding and falling along a road blocked with masonry, eyes empty, mouth slightly open, head shaking from side to side — three days after the earthquake.

Minor earthquakes are something which you surprisingly rapidly learn to live with. You learn not to put precious ornaments near the edges of shelves, where they can be shaken off; you learn that a building is astonishingly forgiving of an earthquake, whether it is a wood-built house which creaks and rearranges itself, or a steel-reinforced concrete skyscraper which rocks and rolls, but never really suffers much. You learn to build on concrete 'rafts' for foundations, and not to live at the top (or the bottom) of steep slopes. You learn to live with the inconvenience of a few hours' disconnection from mains services.

A major earthquake is something completely different. Although for a few moments it feels much like a smaller tremor, it rapidly builds up into an uncontrolled and uncontrollable shaking, shattering force. At the time, it is as if a giant hand is shaking everything brutally, as a bully might shake a child. You cannot walk: you stagger as the ground rises and falls beneath your feet. If you are driving a car, you swerve helplessly as the earth shakes from side to side as well as up and down. There are endless paradoxes: some tall buildings wave like corn in the breeze, but do not fall, while other more solid-looking single-floor buildings collapse in a shower of masonry and dust. You can be standing yards from the nearest building, yet lose your footing, fall against a rock, and die; or you can be in a huge building, and providentially survive as it collapses around you.

The effects of the earthquake that hit Mexico City were made all the worse by the soft foundations; the whole dry lake bed shook like a jelly, so that the damage was actually worse there than in coastal resorts (such as Acapulco) which were nearer the offshore epicentre. In the first four minutes, it was estimated that 250 buildings were completely or substantially destroyed, and that at least 1000 more were rendered unsafe; and when a second shock, much weaker than the first, occurred 36 hours later it demolished several of the ones which were weakened. Many of the buildings were not built to the rigorous 'earthquake-proofing' standards which are applied in, say, California; worse still, the 'bowl of jelly' effect meant that some could resonate to the frequency of the earthquake, shattering themselves to bits like the famous Tacoma Narrows bridge as they shook in tune with the earthquake, vibrating like a child's twanged ruler. In any case, the duration of the 'quake — four minutes is a very long time for a tremor of this magnitude — meant that even 'earthquake-proofed' buildings had plenty of time to resonate, a factor not normally allowed for in building standards for earthquake resistance.

Normally, everything is for sale in Mexico City. You can buy Japanese and American luxuries; you can buy a shoeshine; you can buy a whore (including a twelve-year-old, if that is what you want); you can buy cheap gasoline; you can buy a policeman, because the Mexican police are notoriously corrupt. But after the earthquake, no-one thought of money or personal gain. The police worked like heroes; the armed services pitched in; and ordinary people who were lucky enough to have survived the 'quake worked like men and women possessed. Those who had not lost their houses offered shelter, food, and bedding to those who had lost everything; paradoxically again, it was in many ways Mexico's finest hour, except perhaps for the government, which was widely criticised for indecision and inaction.

Other nations responded as heroically as the Mexicans themselves. Across the American border, their giant neighbour immediately began to feed in aid of all kinds: one of the first teams to arrive consisted of demolition experts who could assess which buildings were safe to attempt rescues, and which were not — and who could help to make the agonising decision to demolish the rest of the building, in the full knowledge that it might mean the end of hope for anyone who might be trapped in there. The French sent in a rescue unit who became known as *Les Taupes*, 'the moles', as result of their skill at digging survivors out of the rubble. From Britain came military and police teams skilled in rescue work, with highly trained sniffer dogs to help find traces of anyone who might be buried. As a result, people were still being rescued as much as two weeks after the initial 'quake, though hundreds or even thousands must have died an agonising death of hunger, thirst, and loss of blood during those two weeks.

Water and power supplies had to be jury-rigged as quickly as possible, but there was the awful possibility of cholera if the water became infected: the routine advice after an earthquake, even if the mains water appears to be coming through properly, is to boil all water before drinking it. It was also a major priority to re-establish the telephone system, which may seem strange until you realise that a functioning telephone system makes it vastly easier to coordinate the immense task of rescue and reconstruction; this task fell to British army volunteers, who had to try to rescue anyone they could from the wreckage of the telephone exchange while they were trying to rebuild it.

Now: suppose that the Mexico City 'quake had been what Californians call 'the big one', and that the San Andreas Fault had ripped free that day. What would have happened?

The answer to that question is like the answer to so many questions in this book: nobody knows. The chances are that proportionately more buildings would survive, because California 'earthquake-proofing' regulations are much more stringent than those in Mexico. But the population of California is now approaching half that of Mexico as a whole, and certainly exceeds the population of the Mexico City conurbation. A lot would depend on when the 'quake happened: anyone who has seen the freeways of Los Angeles at rush hour, jammed solid where they funnel though the city at around Third Street, can only shudder at the thought of what might happen if it struck at (say) 4:30 in the afternoon, when the offices were still mostly occupied, but the freeway was jammed. How could emergency vehicles get through? The answer there is easy: they couldn't, not without a bulldozer-equipped tank to clear a path. Nor is the immediate destruction of the city the only problem: fortunately, there are very few oil wells near the centre of LA, but in some areas there might still be enough to start a fire-storm of the type which destroyed Dresden and Hamburg.

What is more, there was no real *tsunami* (tidal wave) associated with the Mexico City 'quake. An epicentre further offshore would probably have generated a far worse wave; as it was, the wave was about two feet high in El Salvador, 800 miles away, and apparently undetectable in Hawaii, some 3500 miles away. A tidal wave off the crowded California coast could be disastrous — though it could not, as described by the overheated imagination of one writer, 'reach as far inland as Nevada'!

Nor is California the only possible target. Tokyo (and the whole of Japan, for that matter) is extremely vulnerable, and four of the worst 'quakes in this century (measured in terms of death toll) have been in China, where about three quarters of a million people have died in the great earthquakes of 1920, 1927, 1932, and 1976; these are official figures, and actual figures may be much higher, as China's official figure of 242,000 for the 1976 'quake (rather more than a year after Comet Kohoutek) has been criticised by some as being maybe *three times* too low, and the true figure may be nearer three quarters of a million *for that 'quake alone.* Italy has been hard-hit, too: about 75,000 died in the 1908 earthquake (about 16 months before the last perihelion of Halley's Comet), and 30,000 in 1915 (shortly after the 1914 apparition of Delavan's Comet).

By late October 1985, the Mexico City earthquake was the only really major earthquake disaster within our 18-month period — it actually occurred less than five months before perihelion — but others were by no means impossible, and there were reports of a 'quake in Russian Central Asia. Rather less drastic, but still furnishing a great deal of food for thought, were the extraordinary weather patterns which preceded Halley's Comet's 1986 apparition.

To begin with Europe, there was the extremely harsh winter of 1984/85, when even the normally balmy Cote D'Azur in the South of France was covered in snow. This is one place where figures really do tell the story: the Cote D'Azur hit −12°C (10°F), and Florence in Northern Italy reached −21.4°C (−6°F), but the coldest temperature in Germany was −37°C (−35°F), and in Switzerland it was −41°C (about −42°F); by comparison, Moscow was a mild −20°C (−4°F), and Iceland was a positively balmy 8°C (49°F), well above the seasonal norm of −27°C (−16°F). In Britain, just to add to the fun, −17°C (10°F) combined with 100 mph winds to produce truly staggering blizzards!

These weather patterns may not mean much to hardy North Americans, especially New Englanders who regularly expect winters in the 0−20° F (say −10° to −20°C) range, but, to the average European, they were nothing short of astonishing; even the Pope added a special blessing for skiers to his regular benediction! In Britain particularly, where a couple of days below freezing reduces much of the country to a complete shambles, and where insulation standards are among the lowest of any cold, northern country in the world, the impact in the worst-hit areas was devastating, with livestock freezing to death, roads blocked, and railways at a standstill. The British Isles had suffered a record-breaking winter in 1981/82, but the rest of Europe had experienced nothing like it for decades, or, sometimes, nothing like it within living memory.

Not only that; there were actually *two* cold spells, one on the heels of the other, and although the first cold snap in January was bitter, it only lasted for a week or ten days. A couple of weeks later, it froze all over again — enough for the almost legendary Friesian Eleven Cities Tour, a 124-mile (200 km) marathon across the frozen canals and lakes of northern Holland, the first time the race had been run since 1963. Queen Beatrix of the Netherlands awarded the winner's wreath to Evert van Bentham, who covered the distance in 6 hrs 46 mins 47 sec. I was in the South of France at the time of this second cold spell, in late February, and although the snow had disappeared from Cannes (and indeed it was shirt-sleeve weather in Arles), I was trapped by ice, snow, and freezing rain on the way back, while travelling through the Loire Valley, traditionally one of the most sheltered areas of central France!

Meanwhile, in January on the other side of the world, bush fires were blazing in Australia under mile-a-minute winds and 100°F (38°C) temperatures which dried everything to tinder; dozens of homes were destroyed, and many were engulfed by the flames. In a week or so of fires, over $30,000,000 worth of damage was done, including the destruction of a dozen homes in Melton, a dormitory town only twenty miles from Melbourne. It is only fair to point out that the damage was nothing like as great as in the Great Dry of 1983, when over 2000 houses were destroyed and 72 people killed, but, equally, the loss of life could have been much greater if people had not taken evacuation orders very seriously indeed. If there are no serious bush fires for another few years, complacency may take over again — but if the comet were in any way implicated, we might expect a repeat in the summer (ie December/January) of 1985/6....

And then, of course, there was the Great British Summer.

Jokes about the weather in Britain are commonplace, and many visitors wonder whether the British ever talk about anything else; but the rainfall was incessant, like a monsoon in the Himalayas. It was not actually the wettest summer since records began — that dubious distinction belongs to 1927 — but it certainly seemed like it. Looking back on the so-called summer in the surprisingly mild and dry early autumn, people were saying things like, 'It became a way of life; if you looked out of the window and it wasn't raining, you looked again — and it was.' It was the topic of endless conversations and speculations; some wondered if it would ever stop. Even the bookmakers began to have their doubts: after offering odds of 25-1 that it would not rain for the whole of August at the beginning of the month, they had dropped to 2:1 by the end of the month, although in the event they did not have to pay out, because there was a brief respite and a little sunshine for the August Bank Holiday, a fact for which the insurance companies who had underwitten pluvius (rain) insurance were profoundly grateful.

The farmers suffered worst, with the British government making it clear in October that they were going to provide some sort of financial aid; the Irish, on the other hand, were talking about applying for $30,000,000 from the European Economic Community's disaster fund. British farmers lost 15-20% of their wheat crop, and root crops rotted in the ground, attacked by insects and disease. The abnormal weather was not over, even at the time of writing: in Britain at least, October was the warmest it had been for half a century, and a representative of a major oil company took the possibility of comet-induced weather changes seriously enough to call the Royal Astronomical Society in order to get their views; a warm winter would, of course, have a bearing on the demand for oil. A spokesman of the Society gave him a somewhat dusty answer, but when multi-million pound corporations take things this seriously it is no surprise that the man in the street wonders if there is any truth in it! At worst, however, the British farmers' loss was only financial, though the word 'only' may bring a bitter reaction from those who did lose money. What I mean, though, is that there was neither starvation nor massive loss of life from the weather; that was reserved for other countries.

In Bangladesh, June brought a *gurnijhar* or 'swirling wind' — in Atlantic parlance, a hurricane, or in Pacific terms, a typhoon. Because of the unique geography of the Bay of Bengal, the force of any hurricane is funnelled into the Ganges delta, and invariably brings widespread flooding and loss of life. At least 15,000 were killed in the desperately poor country which used to be East Pakistan; the water came in like a wall, faster than a man could run, and thousands were overtaken even as they were running for shelter to the few concrete-built buildings which would not be washed away. The June disaster was the worst since 1970, when an estimated 300,000-500,000 people were killed; that *gurnijhar* coincided with Comet Bennett . . . .

In the Sudan, virtually rain-free for seven years, and in the grip of a bitter famine, rain fell in such quantities in July that the hard-baked earth simply could not absorb it. The result was flash floods, walls of water tearing along gullys and dry river-beds, and ripping up and smashing anything that got in their way — people, vehicles, roads, and railways. Most of the roads in the Sudan are dirt, and the wheels of trucks bearing relief supplies from Port Sudan into the interior rapidly churned these into a quagmire, so that they became impassable; the same happened to airstrips. Thousands of tons of food from the world's relief organisations began to build up into macabre stockpiles, in a country where 30,000 people were expected to die of hunger *every month*.

In Italy, July thunderstorms caused the failure of two dams in Stava, a resort and fluorite extraction centre in the Dolomites about 130 miles north-east of Milan. Only eight people were found alive; between 200 and 250 died. By Sudanese or Bengali standards, or by the standards of the Mexico City earthquake, it did not amount to much; but every single one of those 200 or more who died was an individual, directly comparable with an individual dying in Mexico City.

In the United States, too, there were hurricanes. On 27th September, Hurricane Gloria rampaged along the Atlantic coast: because of good planning and rigorous safety measures, loss of life was very slight, but property damage ran into millions — despite such precautions as Macy's, the New York department store, closing *and boarding up its windows* during the hurricane's passage. The earlier hurricanes (Hurricane Elena) which ravaged the states surrounding the Gulf of Mexico in mid-August fortunately struck hardest in thinly populated areas: disastrous for the poor farmers and fishermen trying to grub a living in some of America's poorest states, but, on a national scale, mercifully slight. On the one hand, a country like America has several advantages when it comes to natural disasters: the various State and Federal authorities have contingency plans, radio and television are almost universally available as a means of warning people, who can then leave the area by personal or public transport, and high building standards mean that the damage is far less than in, say, Bengal. There, it is all but impossible to provide adequate warning, because of the lack of communications, and because people not only have nowhere to go, and no means of getting there, but face the almost certain destruction of all they possess. On the other hand, the very high standard of living of the United States means that even 'superficial' damage is severe in financial terms: smashed glass, domestic appliances and carpets ruined by mud and water, cars destroyed, mains services disrupted, even the mass of paper and other information (for example, on magnetic tape), all have to be taken into account when assessing damages. The fact that the insurance will (eventually) pay for it does not make coming back to this sort of scene in your own home any more attractive.

The list can be continued. Also in August, Typhoon Nelson killed at least twenty in China — probably more, because Chinese authorities almost always understate the level of casualties, as though admitting that someone had died was an admission of the failure of Communism. A couple of weeks previously, ten or more were killed in Korea by Typhoon Kit; a couple of weeks later, seventeen were killed and hundreds injured in Tokyo by Typhoon Pat. At the time of writing, the typhoon or hurricane season (the two words describe the same phenomenon) was generally reckoned to be over, but the 1986 season will begin three or four months after the perihelion of Halley's Comet and last until seven months after; the 1985 season began almost ten months before and ended five months before. The 1986 season is unquestionably closer to perihelion....

Was all of this asssociated with Halley's Comet? If it was, no-one knows how, but the factors which control the weather planet-wide are *very* imperfectly understood. It is hard to imagine that a system which can command such awesome forces is so delicately balanced that it could be affected by the minute variations in gravity, electromagnetic field, or solar wind which the comet would cause, but we have no better explanation: certainly, forces on this scale, and this far beyond man's control, fall well outside the realm of self-fulfilling prophecy (at least without getting *very* mystical and metaphysical about the whole thing), and Jung's theory of synchronicity does not seem to have very much to offer, except in the astrological field. When you consider that a single typhoon of the magnitude which struck the Bay of Bengal dissipates the same energy as 150,000 *megatons* of nuclear bombs, albeit over a period of days rather than seconds, you realise what a titanic scale the world operates on; can it be that there are equally gigantic forces *of which we know nothing at all* which control the world weather? If so, what controls them?

6
BLOOD ON THE STREETS

The deliberately emotive title of this chapter refers to three separate threads which have — literally or figuratively — put 'blood on the streets' of cities all over the world. These are riots; terrorism; and disasters such as Bhopal, where human error combined with chance and nature in a way that is mercifully rare, but which was (as is the thesis of this book) all too common as the world waited for Halley's Comet.

With all three strands, it is important to distinguish between the setting for the event, and the trigger. We live in an unstable world, and one in which disaster can strike at a moment's notice, or with no warning at all: this has been true of most of the twentieth century. But as the world grows more complex, so does the scope for disaster grow. The question is, what triggers the disaster?

Sometimes, the answer is clear: the political background to the assassination of Indira Gandhi, for example, is fairly easy to understand. Sometimes, we seem to be back to 'pure' chance again, as with the Bhopal chemical plant disaster or the liquefied-gas explosion in Mexico in November 1984. Often, though, the answer lies somewhere in between. There will be an ostensible answer, an individual's action (such as the police shooting which triggered Britain's Brixton riots in September), but it will not be a complete answer: it will be more like an excuse than a reason. The newspapers are always ready to apply their clichés, with words like 'tinderbox' or 'explosive situation' awaiting the 'spark' that sets them off — but these words answer nothing. Why did an area become a 'tinderbox' in the first place, and why was it *that* spark which set it off? The riots in Handsworth in Birmingham, for example, were apparently triggered by a perfectly legitimate questioning of a driver of whom the police had good reason to be suspicious (the tax disc on his car was out of date); how does that justify a riot which did millions of pounds' worth of damage and left several people dead? And what determines the degree of violence in a riot? Why was a policeman hacked to death with a machete, and his head almost severed, in the Tottenham riots when in the Bristol riots a few years previously there had been hardly any serious injuries?

If we begin by looking at riots, we can see that there are obviously some areas in which rioting was all but inevitable; in South Africa, for example, tension between blacks and whites has been so great for so long that no-one was really surprised when open violence erupted. Certainly, matters were not helped by the snap judgements of the media, some of whose correspondents (one suspects) had only the vaguest idea of where South Africa was before the trouble started, but it would do little good here to go into the history and politics of that troubled country. Similarly, the Jamaican riots of January 1985 were primarily over price rises, and the continuing trouble in Poland is due as much to the fact that it is a nation occupied by an oppressive foreign power as to any specific events. A much more fruitful field for study is the various British riots and (especially) the football riots, many of which involved British fans; the Philadelphia Move raid, where the force involved was apparently so far out of proportion to the threat as to be incredible, is also worth investigating.

There was a foretaste of what was to come in Britain at the annual Notting Hill carnival, which has been the scene of many scuffles between police and rioters, when a 51-year-old Filipino man was beaten to death by what was described at the time as a 'small army', but the first wholesale outbreak of lawlessness began in Handsworth in Birmingham on Monday 7th September. As already described, a policeman stopped a driver in order to question him about an expired road fund licence; this was at 5:30pm on Monday. Rapidly, a crowd of about a hundred youths gathered around the policeman, taunting him and telling him to let the driver go. He radioed for help; two more policemen arrived; there was a scuffle, in which two black youths were arrested; the crowd dispersed; and that seemed an end to it.

Later that evening, at 7:45, a fire was reported in a derelict bingo hall, but when the fire department came to extinguish the blaze, they were met with bricks, bottles, and even Molotov cocktails: on Wednesday, many of the papers were to carry the same picture of a young black, flaming Molotov cocktail in hand, waiting to throw it. The firemen called the police; within minutes, several hundred rioters were running down the Lozells Road, a seedy but hitherto safe part of Handsworth. What happened next is incredibly confused. The core of original rioters were rapidly swelled by looters, some of whom had apparently seen the riots on television and driven over post-haste to join it; this may seem astonishing to most readers, but in the Bristol riots half a decade before I learned that some of my friends had gone down 'to watch', though whether it was fear of the consequences or a genuine wish not to break the law which kept them from joining in, I never discovered. Vehicles and buildings were burned: in one post office, two men were burned to death. One Indian grocer was robbed by hooligans who drove a car straight through his shop window, demanded all his money (and threatened to burn the place down if he did not give it to them), and then stripped his shelves. Police reinforcements were drafted in as fast as possible; by 9 pm, the police were able to advance over the burning barricades, and the area was sealed off by 10:30 pm, with about 600 police on the scene. Sporadic rioting broke out again next day, when the Home Secretary went to visit the area; final estimates of the cost of the riot were in the millions, though the death toll rose no higher.

In the course of my life, I have been close to three riots; I was in Malta during independence riots in the late 1950s, in Bermuda during the independence riots of the late 1960s, and in Bristol during the St Pauls riots of the early 1980s. The two things which struck me and which seem characteristic during a riot are, first, the violent swings of mood, and second, the remarkably limited area over which most riots take place; the first I knew of the Bristol riots, which were taking place about a mile away, was when my wife's parents called to ask if we were all right, and when we asked why we should not be advised us to turn on the television set!

During a riot, there is a sort of reckless party atmosphere, when reality seems suspended, but when suddenly you are on the receiving end of the violence you realise how terrifying and arbitrary it is. When I was younger, the father of a girlfriend was deliberately run down by a car; it was certainly no accident, as they reversed over him to be sure he was dead.... I have twice been in a car that was being stoned, though both times the stoning was fairly desultory and we managed to get away, and in Bermuda my father was riding a moped (a standard form of transport on an island where private cars are limited to one per household) when he glimpsed a flash of sunlight from a piece of piano wire strung across the road; if he had not seen it, and stopped, he would have been decapitated.

The Handsworth riot exhibited just this sort of arbitrariness, but it lacked the kind of underlying feeling which had obviously triggered the Malta and Bermuda riots. The 'official' explanation (by which I mean the one most widely accepted, especially by the government) was that the initial rioting was due to black resentment against the police, because Handsworth contains many blacks and is a centre of drug-dealing, while the second wave was due to sheer opportunism. While the phrase 'drug-crazed' will bring a wry smile to the lips of most who are familiar with the effects of drugs — with a few exceptions, such as 'Angel Dust', most drugs leave their users listless and apathetic — it is also true that drug users can often be very easily manipulated, especially when they are stoned, and that drug dealers have good cause to want to keep the police away.

It was the second wave of rioting — the opportunism — which is in a way more disturbing, because it does seem to reflect a general disrespect for, or disregard of, law and order. You cannot blame it on racial tensions, because the rioters were both black and white; you cannot really blame it on unemployment, because (for example) unemployment in Cornwall is very high, but there are no riots in Truro or St

Austell. You can blame it on the environment to a certain extent — I attended Birmingham University, and lived for a while in Balsall Heath, another rough part of the city, and I know that inner-city Birmingham is pretty dire — but we had no riots in those days. The media used to blame riots on 'long hot summers of discontent', but the British summer of 1985 blew that excuse; besides, the 'long waterlogged summer of discontent' will not fit so easily on a headline, so a more serious attempt has to be made to divine the cause of the trouble.

Less than three weeks later, there were riots in Brixton — another traditional trouble spot, though one that had been quiet for quite a while. There, the trouble arose when the police accidentally shot a black woman while conducting a dawn raid on Saturday 28th September in connection with alleged firearms offences by her son, who was subsequently charged. There seems little doubt that the unfortunate woman in question, Cherry Groce, was injured accidentally, but during the afternoon, while the police were searching the house *and investigating the shooting*, a crowd gathered outside and began to hurl insults and (eventually) bricks; a bottle was thrown through the windshield of a policeman's car at 4:30 pm, and by 5:30 there was an angry, stone-throwing crowd around the police station. What followed was a smaller-scale replay of the Handsworth riot, with looting, burning, and an immense amount of opportunist crime, including two separate rapes of white women; over 500 separate charges were brought, though serious injuries were mercifully small.

A few days later again, on 1st October, there were riots in Toxteth, in Liverpool, when four young men were arrested in connection with a murder inquiry; four vehicles were turned over and burned, and there was some looting, but it was to be London's turn again on 6th October.

Once again it was a black woman, Mrs Cynthia Jarrett, who was the innocent victim in the event which sparked off the riot. Police were raiding her house, again looking for someone else, and she suffered a heart attack. The police apparently did not recognise the symptoms, and just kept asking her where the boy was. Although a minor heart attack of the kind which she suffered need not be fatal if it is treated sufficiently rapidly, it was not treated at all, and she died.

Although one can sympathise with the feeling of the rioters in Tottenham as in Brixton, one cannot sympathise with what they did: 254 people were injured, and one — a policeman — was hacked to death with knives and a machete, one blow of which apparently all but severed his head from his body. According to an eyewitness, he tripped and fell, and three or four people attacked him with knives. Three people had been charged at the time of writing, one white and two black: one was a fourteen-year-old boy. Nor can one sympathise with the leader of the local council, who said, 'The police got a good hiding, and they deserved it....'

It was seriously suggested in both the London riots that extremist groups orchestrated much of the violence, including attacks on the police, but there certainly seems to have been a good deal of opportunist crime under the cover of this avowedly 'political' riot, so that it looked to an impartial observer as though the revolutionaries were making unrealistic claims on the one hand, and that these claims were being accepted by the media and (some) politicians on the other, because they provided a convenient explanation for the whole thing! At the time of writing, the Tottenham riots were still simmering down, so it was too close to make any better assessment; but there did not seem to be any relationship between the force used by the rioters and the initial provocation. A traditional English expression of helplessness in the face of such events is to say, 'I don't know what's got into them' — and the more you think about the literal meaning of that observation the more thought-provoking it becomes.

The Move disaster in Philadelphia was almost a mirror image of the British riots; in response to what was admittedly a fairly major threat, it was the turn of the police police to overreact, which they did with the most extraordinary degree of overkill. 'Move' was by its own lights a back-to-nature movement, always assuming that you allow 'nature' to include firearms and explosives, which they constantly threatened to use in order to preserve their lifestyle on Osage Avenue, Philadelphia. Their lifestyle was bizarre in the extreme: they adopted names such as Ramona Africa, and gave their children names like Birdie Africa (to name two of the survivors of the attack), and lived among their own filth, rats and pests with the children going naked and never attending school; they ate raw meat, and foreswore 'artificial' heat. Neighbours report that they also used powerful sound systems or simple electronic bull-horns to preach their beliefs (often in obscene language) at all hours of the day and night, or simply to play music. They physically assaulted some of their neighbours, and threatened others, who were understandably fairly unanimous about wanting them out; Move's reply was, 'If you do anything to hurt us, we'll kill you.' Nine members of the cult were already in gaol as the result of the murder of a policeman in a previous confrontation when the May 1985 problems came to a head.

Police had obtained warrants charging four of the occu-

V & A/Phaidon

Halley's Comet as depicted in the Bayeux Tapestry, recording the disastrous defeat of Harold, King of England, and his armies in 1066.

An actual photograph of the 1910 appearance of Halley's Comet *Mary Evans*

An artist's reconstruction of Halley's Comet over Paris in 1910.

(Right) Police remove a body from the rubble of Mexico City following two earthquakes which hit the city within 36 hours of each other killing many thousands. September 1985.

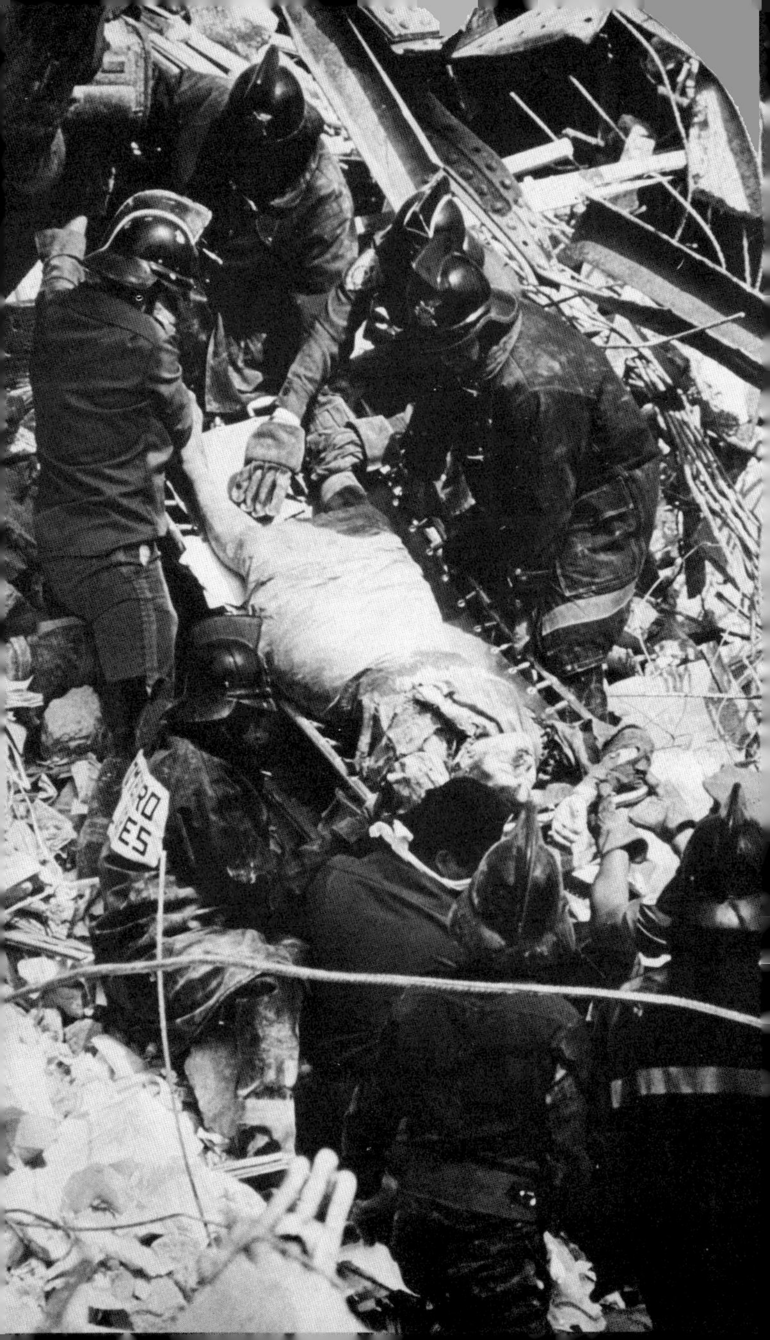

Popperfoto *Popperfoto*

operfoto

agnum

(Top left) The aftermath of destruction of Hurricane Elena at Indian Rocks Beach, Florida. January 1985

(Bottom left) Fans struggle to get clear of a stand wall which collapsed at the Heysel Stadium, Brussels. May 1985.

(Top right) Police in riot gear file past a burning car following an eruption of trouble in Brixton, South London. September 1985.

Masked Shi-ite militia men fight in the streets of Beirut. October 1985.

The remains of the hi-jacked Boeing 737 blown up at Beirut Airport. June 1985.

The Collision of two passenger trains at Alcafache, Portugal which resulted in 60 deaths. September 1985.

Police and rescuers walk around the remains of the Paris-bound express which smashed into a truck at a crossing in Saint-Pierre-du-Vauvray in Northern France. July 1985.

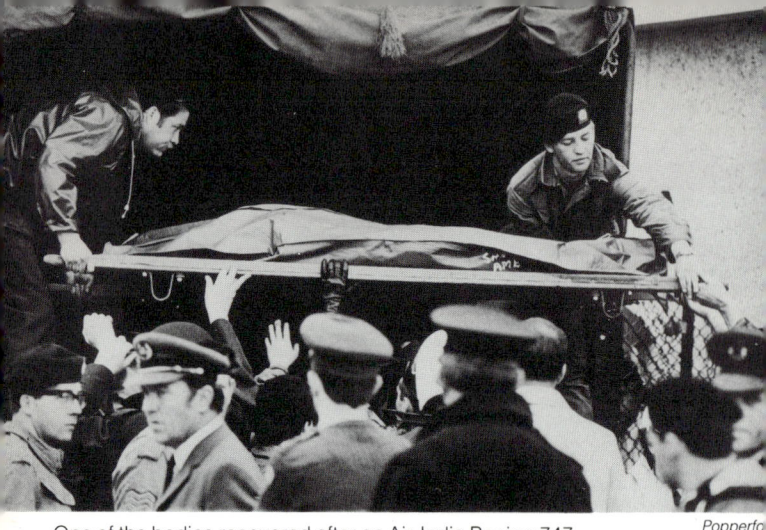

One of the bodies recovered after an Air-India Boeing 747 plummeted into the Atlantic 150 miles off the coast of Ireland. June 1985.

Popperfo

Wreckage of the Japan airlines Boeing 747, which crashed on a ridge near Mount Osutaka 110 km northwest of Tokyo, killing 520 passengers and crew. August 1985.

Popperfo

pants of the house with parole violation, contempt of court, illegal possession of firearms, and making terroristic threats. They anticipated difficulty in executing the warrants, and evacuated 300 neighbours from a cordoned-off area around the house. On the morning of 20th May, there were about 150 men surrounding the house; at 5:35 am, Police Commissioner Sambor announced through a megaphone that he held the warrants, and that the people named in them had fifteen minutes to come out. The Move members inside the house replied with jeers and obscenities. When the quarter hour was up, police threw tear-gas grenades into the building, and the fire department drenched the house with water: Move had threatened to 'blow the block' if they were attacked. A burst of gunfire from inside the house was answered with *an hour and a half* of concentrated fire from the police — though it is only fair to add that whenever they tried to approach the house in any other way they were driven away by gunfire.

With sporadic fire continuing, and the house very much the worse for wear as a result of the thousands of gallons of water that had been poured onto it, the police decided to use aerially delivered explosives — or in other words, a bomb. The explosive used was Du Pont Tovex TR-2, which generates intense heat (3000-7000°F), and which according to the manufacturers is designed for underground mining and quarrying. Despite the force of the explosion, which a neighbour in an adjacent street said shook his entire house, the bunker which Move had constructed in the basement of the house was apparently not destroyed, and the police decided to let the fire burn for a little while in order to flush the Move members out. But fires are not necessarily controllable, and when it began to surge along the houses alongside the wreck of the Move building, firemen found that they were unable to control the blaze, because they were driven back by gunfire. The flames spread from Osage Avenue to Pine Avenue; there was nothing that could be done. The fire department could contain the outskirts of the fire, but they could not get close to its centre. When the fire burned itself out, the police found eleven bodies in the Move building. Four of them were children. They had also destroyed 53 houses, severely damaged eight others, and left about 240 people homeless. The cost of the damage was estimated at $8,000,000.

The question again offers itself: 'What got into them?' What got into the Move members, that they would sacrifice themselves and their children, and what got into the police, that they could wreak destruction on such a scale? Was it a catalogue of misjudgement and misunderstanding — or was everyone affected by some sort of mental state which made this the only possible course of action?

Moving on to football riots, the rowdy behaviour of British fans has long been a cause for concern, though it is important to remember that they are neither the only offenders nor the worst. In 1964, over 300 people were killed, and more than 500 injured, in a fight between Peruvian and Argentinian fans in Lima, Peru. In 1967, 48 were killed and 602 injured in Turkey, and in countries as diverse as Mexico, Calcutta and even China there have been football riots: at the end of May 1985, when Hong Kong knocked China out of the running for the World Cup, 30 policemen were beaten up (four seriously) and 127 arrests made, though fortunately no-one was killed.

Nevertheless, Britain has for most of the twentieth century been a (fairly) law-abiding country, and the rise in football violence throughout the 1970s (since the appearance of Comet Bennett, one is tempted to say) has defied analysis.

The Brussels disaster, when British fans fought Italians before the European Cup Final had even started, was horrific; as a Belgian Red Cross worker put it, 'This is not sport. This is war.' Thirty-eight people died, and as many as 500 were injured, some seriously; an image which is hard to lose is of a young Italian teenage girl, trapped on top of a barbed-wire safety fence, dripping blood, while British fans threw bricks at her. One of the most telling comments came from a British club manager; rather than blaming the carnage on the fans, he said, 'The stadium was totally inadequate; there was no way that it could get a safety certificate in this country' (ie Britain). Does he mean that he expected mass murder? If so, one can only ask of him, as well as of the fans, 'What has got into them?'

Terrorism poses equally as many puzzles. It has been around for a long time; no doubt the leaders of the American revolution would have been called 'terrorists' if the word had been in common use in the 1770s, the Russian nihilists of the last part of the nineteenth century were 'terrorists', and so was Gavriel Prinszep, the man who shot Archduke Ferdinand in Sarajevo. Terrorism as a form of *random* attack, however, was little used until the 1960s — though the IRA had made some moves in that direction at the time of the Irish independence struggle. What is strange is that there should, after a long period of relative quiescence, have been such an outbreak of 'high-profile' terrorism.

Aircraft hijackings had shown a steady decline since the early 1970s until 1985, when there were as many incidents in the first six months as there had been in the whole of 1984 — and the only really dramatic hijacking of 1984 was at the very end of the year, on 4th December, when a Kuwaiti airliner was seized and diverted to Teheran by Shi'ite extremists who eventually killed two American passengers simply because they were American; this hijacking came to an end when Iranian forces stormed the aircraft.

A period of relative quiet followed — by which I mean the dozen or so hijackings which occurred were all resolved without casualties — until June, when there were three hijackings within a week, though only one was to have fatal consequences. The first was of a Jordanian 727 on 11th June, when six more Shi'ites gained control of the plane at Beirut Airport, ordered the pilot to fly to Cyprus, and then pursued a sort of aerial Flying Dutchman's course around the Eastern Mediterranean, with no-one willing to give them shelter, until they ended up back in Beirut. This hijacking ended when the terrorists released the passengers, blew up the aircraft, and disappeared into Shi'ite enclaves near the airport where they could be sure of shelter. For some passengers, there was an ironic and terrifying sequence to this; they caught the first available flight out of Beirut, on a Middle East Airlines 707, but when it landed on schedule in Cyprus a Palestinian terrorist produced a grenade and demanded to be flown to Amman. Fortunately, he surrendered to the captain when he was told that he would be flown there on a Jordanian airliner. The Palestinians are Sunni Moslems, the majority sect, against whom the Shi'ites are struggling....

Then, on 14th June, the ill-fated TWA flight 847 was hijacked out of Athens, diverted to Beirut (where the airport authorities were very unwilling to let it land), flown to Algiers, then back to Beirut; there, just to prove that they were serious about their demands for the release of 776 Lebanese held in Israeli gaols since early 1985, they shot a young man whom they said was a US marine, and dumped his body on the runway. At last, their requests *were* granted — and they released the hostages.

One of the most alarming aspects of this last hijacking is that it reveals the disquieting spectre of official or semi-official terrorism; the Israeli-held Lebanese were themselves a form of hostage, taken by Israeli forces during their withdrawal from Lebanon as a deterrent against Lebanese counter-attacks. In fact, over most of 1985, Israel's public image suffered badly, while that of the Palestine Liberation Organisation in general, and Yasser Arafat in particular, improved: it is easy to forget that Menachim Begin was once a 'terrorist', occupying much the same position as Yasser Arafat in the eyes of the British during the setting up of Israel in the wake of the Nazi holocaust. Intransigence is forced upon him by circumstance — Israel simply cannot afford to make concessions which would affect its borders — but it does give Arafat a chance to look like a reasonable man, because he can offer (apparent) concessions without the slightest risk that the Israelis will accept his offers.

This trend was somewhat reversed in early October when a PLO faction (albeit apparently without the sanction of Arafat) hijacked an Italian ship, the *Achille Lauro*, to publicise the Palestinian cause and (once again) shot a single passenger as earnest of their seriousness. This hijacking came to an end when the terrorists were given an airliner to fly them home, but their plans went awry when American fighter aircraft forced their plane down in Sicily; at the time of writing, they were facing criminal charges (including piracy and murder) in an Italian court.

Much clearer instances of government-inspired (or at least, government-approved) terrorism have occurred in Nicaragua, Sri Lanka, Nigeria and elsewhere. There are also plenty of other examples of 'pure' terrorism, such as the Brighton bombing in October 1984 when a serious attempt was made to wipe out the Conservative party conference, and the assassination of Indira Gandhi on 31st October of the same year. The former was inevitably linked to the Irish Republican Army, while the latter had been seen as inevitable by any serious India-watchers after the Indian prime minister had ordered the Army to go into the Golden Temple in Amritsar after Sant Bhindranwale, a Sikh religious fanatic who directed terrorism aimed at the creation of an independent Sikh state, Khalistan. Even the Russians, who have so often backed terrorist movements (of course, one man's 'terrorist' is another man's 'freedom fighter') were not safe: four Russians were kidnapped in West Beirut on the last day of September 1985, and at least two of them had been shot 'as a warning' at the time of writing.

Finally, at least in this chapter, there are three other major man-made disasters which put 'blood on the streets'. Two have a certain amount in common, while the third stands alone. They are the Mexican liquefied-gas explosion and the Bhopal chemical disaster, both in December 1984, and the Bradford football stadium fire in May 1985.

The Mexican disaster was in a suburb of Mexico City — a city which was, of course, to experience a devastating earthquake a few months later. Just before dawn, something touched off a fire in the San Juan Ixhuatapec gas-distribution centre of the Mexican state petroleum company, PEMEX. PEMEX says that the fire started outside their depot, but are unable to show how that might have happened — and are, of course, unable to counter allegations that even if this were so, the fire should still not have spread.

The scale of what followed was incredible. Four gigantic spherical tanks, each containing over a third of a million (imperial) gallons of liquefied gas, were among the first to blow: they went off like bombs, scattering fragments of their pressure-resistant casings like shrapnel. Then the smaller tanks began to go up in the general inferno: in one incredible incident, a 50-foot propane cylinder was hurled half a mile, falling from the sky like a bomb and completely demolishing a house. The furnace-like heat was bad enough — the pressurised gas heated the whole area so badly that survivors say they burned their feet as they ran away — but the endless explosions made it impossible for firefighters even to approach the centre of the fire, and it was an hour and three quarters before they were able to get close enough for rescue workers to try to get in. When they did, they found literally hundreds of bodies. Some, the lucky ones, had died instantly, often in the explosions; others had roasted slowly to death behind walls, or died of suffocation as the fire used up all the available oxygen in the air. Many were so badly burned that there was no question of identifying them; some may have been incinerated so completely that there was nothing left but a handful of ash, while the only remains of others were a charred leg or hand.

When the flames had cooled, the official estimate was that 365 were dead, 2000 injured. Within a week, as many as a third of the injured were also dead; because no-one knows the population of the teeming shanty-towns which surround Mexico City, the death toll is likely to have exceeded 1000. As we have seen so often in this book, no-one knows how it happened. It was just another disaster which coincided with the comet . . . .

Less than a month later, on 3rd December, there was another disaster beside which the San Juan Ixhuatapec explosions almost paled. It took place in a city in central India of which few had heard before the catastrophe, but which became a household name: Bhopal.

A Union Carbide chemical plant was built there in 1977. It was designed to manufacture the pesticides which India so desperately needs, and inevitably it handled a fair number of lethal chemicals: methyl isocyanate, an essential ingredient of certain pesticides, was only one of them.

It was actually built a fair way out of Bhopal, precisely because there was some risk involved, but because there was work at the plant a shanty town inevitably sprang up around it; the constant inflow of desperately poor Indians from the towns to the cities means that there are always more people who want to work than there is work available, and the man who is nearest the job can get in earliest in the morning. Besides, if home is a shanty made from old doors, packing cases, cardboard, plastic sheeting, and leaves, moving house is no great difficulty; and there was the added attraction that water and power supplies were laid on for the factory, supplies into which the shanty-dwellers could tap.

The colourless, substantially odourless gas which caused the tragedy reacts rapidly and horrifically with water in the human body. Any moist tissues swell and become inflamed, so that the first symptom is a choking feeling as the throat closes up; if sufficient gas is inhaled, the lungs also swell and clog and begin to fill with water, so that the victim literally drowns in his or her own body fluids.

Nor are these the only effects. The eyes rapidly begin to water, and longer exposure can cause cataracts. Liver damage is another proven effect. Children died in the womb after the Bhopal disaster, and doctors had the grisly task of aborting babies who were dead anyway. Beyond this, much is conjecture: the optimists say that isocyanates can be excreted over a period of time, even to the extent of reversing minor cataract damage, while the pessimists point to possible life-long asthma, bronchitis, or emphysema from the lung damage together with blindness, kidney damage, and possibly even brain damage.

People are still dying from the effects of the Bhopal disaster, and the total number of deaths directly attributable to the leak will never be known accurately, but the figure is *at least* 5000, and may well be higher. As with the Mexican earthquake, numbers mean little; the picture of a child's face, peeping from the earth just before it is finally covered by the grave-digger's shovels, means far more. So do the images of pi-dogs dragging bodies from too-shallow graves and eating them; so do the vultures eating the unburied dead. The swollen bodies of cattle, about to burst from putrefaction, littering the fields, tell their own story; the absence of bird-song, even of the chirping or crickets, continues the tale. But what happened?

Unlike the San Juan Ixhuatapec disaster, we have quite a number of insights into the way that the Bhopal disaster happened — and the catalogue of coincidence, misfortune, and misjudgement is as tragic as any.

To begin with, there should never have been as much methyl isocyanate stored in one place. Three large underground tanks stored several *tons* of gas, whereas in Europe it is usual to keep only very small stocks of such toxic ingredients, so that in the event of a failure, the disaster is kept to a minimum. It is also customary to store it in much smaller tanks, each with its own protective measures, so that if there is a failure of any kind, it will be confined to one tank.

Secondly, there may have been some form of 'blow back' into the tanks. Methyl isocyanate boils at about 100°F (39°C), and daytime temperatures in Bhopal often exceed that, so the tanks are refrigerated. Some of the other chemicals also used in the plant react strongly with methyl isocyanate in an *exothermic* or heat-producing reaction. If one of these other chemicals had found its way into a methyl isocyanate tank, the result would have been a rapid heat build-up followed by automatic venting to prevent rupture of the tank.

Thirdly, this heat build-up (and its corresponding pressure build-up) should have been registered far earlier than it was by the warning mechanisms *and something should have been done about it*. As it was, the heat build-up was noticed almost two hours before the gas escape, but the man who noticed it was unable to reverse the effect. The plant had been closed for maintenance for two weeks before the disaster; the panel which would have revealed the problem in the main control-room was disconnected.

Fourthly, in the event of a leak or automatic venting, there is a 'scrubber' or chemical cleaner built into the release vent which should remove small amounts of gas before they can reach the atmosphere; if this fails, or if the volume is too great for the 'scrubber' to handle, the escaping gas should be ignited and burned off as it escaped. The scrubber was overwhelmed, and the burn-off back-up did not work.

Added to this unfortunate sequence of human and mechanical failures, the night was cool and still. One might have expected the tanks to fail during the day, if at all, because of the heat of the sun. If this had happened, the gas would have dispersed much more rapidly, and the disaster would have been on a considerably smaller scale, though still terrible. As it was, the gas spread slowly and hung over a quarter of the city in a deadly cloud.

It is easy to blame the accident on 'the Americans', as though Union Carbide were indifferent to the deaths of their workers, but this is sheer stupidity; the parent company does hold a majority share (actually 50.9%), but the rest of the money is Indian, and the majority of the workers were Indian; besides, no company in its right mind would risk a

$25,000,000 investment, which is what the Bhopal works cost.

It is equally easy to blame 'the Indians', as though the disaster could be attributed to shiftless Third World ways. This is equally pointless. The men at the top were skilled and highly trained; they knew what the results of a leak would be, and they had no more wish to die than anyone else.

The truth, as so often in this book, comes back to 'bad luck'. Safety standards may not have been quite as high as they would have been in the United States, and they may not have been quite as rigorously applied, but they were still pretty rigorous. The chain of misfortunes and coincidences which culminated in the release of the deadly cloud may have been preventable with the benefit of hindsight, but it was not prevented. All we can do is to add it to our list, and to add that on 11th April 1985, the parent company announced quietly that the plant would not be re-opening.

The most recent of these man-made disasters was the fire at the Bradford football stadium on 11th May 1985. The cause of the fire is still disputed — it may have been a cigarette end, or a prank which got tragically out of hand — but the result was that rubbish under the stand caught fire, and the main stand itself then caught fire only seconds afterwards; paradoxically, in the light of the British summer, the old wood was tinder-dry, and the whole stand was ablaze in about four minutes, with rafters and burning debris raining down on fans and would-be rescuers alike. The wind was from the north, the worst possible direction, as it created a force-fed fire; yet another example of nature conspiring with fate to create disaster.

There were about 11,000 people at the match, 3000 of them in the stand. In the panic that followed, 53 people lost their lives, and others were permanently injured. Many were crushed to death, a peculiarly horrible way to die which can take ten minutes or more and leaves the corpse purple from lack of oxygen: is is essentially a slow form of asphyxiation. Others were covered with molten asphalt which ran from the burning roof; it solidified around them, so that police had to chip and pry the corpses loose.

There are three details which make the story all the more terrible. First, the club had been warned by West Yorkshire County Council that the main stand was unsafe — but the letter had apparently gone unread because the club had gone bankrupt and changed owners, and mail was still being diverted to the receivers. Second, the gates were locked, to prevent latecomers from sneaking in without paying, but there was no form of 'crash lock'; more than one door had to be burst open by panicking survivors fleeing the inferno. Third, fire extinguishers had been removed because fans had been using them as missiles, or letting them off in the faces of rival supporters; thus, the few fans who did not respect law or commonsense effectively murdered those who did.

The Bradford fire was not on the same scale as the Mexican fire or the Bhopal disaster, but every individual who died was just that: an individual. Can we link Bradford with the comet? There is no way we can prove any direct link; we can just add it to our list.

7
PILOT ERROR?

Airline crashes are a fact of life; they are something which we all know about when we buy an airline ticket, or when we board an aircraft. We also know that, statistically, we should worry more about being knocked down in the street by a car than about being in an air crash. The same is even more true of rail accidents, to which we shall return at the end of the chapter.

Although the odds are normally such that we can ignore them, the record of air crashes in 1985 was astonishing. At the time of writing (mid-October 1985), there had been six major crashes (defined as crashes in which more than fifty people are killed), and there was at least one *very* near miss; at the end of February, China Airlines Flight 006 from Taipei to Los Angeles fell six miles through the sky in two minutes. The China Airlines 747 was lightly loaded, with only 243 passsengers and 25 crew on board; when one engine failed, and the other three reduced power, the 325-

ton plane rolled to the right and began to fall at nearly three miles a minute. That everyone on board was not killed is due to the skill, experience, and superhuman effort of the pilot; to the fact that they were flying at 41,000 feet; and to what the pilot, Ming Yuan Ho, described as the 'miraculous' way in which the engines regained power at about 11,000 feet, allowing him to regain control at about 9000 feet (about 44 seconds before the aircraft would have crashed) and to fly across 350 miles of open sea for an emergency landing at the nearest airport, San Francisco.

Had CA 006 crashed, it would have been the third crash of 1985. About a month before a Galaxy Airlines Lockheed Electra jet-prop had crashed in Nevada, killing 68 of its 71 passengers, and a mere ten days before, an Iberia 727 had flown into a mountainside in Spain, killing all 148 on board. Under four months later, an Air India 747 disintegrated in mid-air over the sea off the coast of Ireland, with the loss of all 329 aboard; five weeks after that, a Delta Airlines L1011 crashed while landing at Dallas/Fort Worth; ten days later again, there was the worst single-aircraft disaster in the history of aviation, when a Japan Airlines 747 smashed into a mountainside in a remote area of Japan, killing 520 people out of 524; and ten days later again, in Manchester, a British Airtours 737 was destroyed by an engine explosion during take-off, and 54 of the 137 passengers were killed. These six crashes were only the *major* ones; by the beginning of September, there had been fifteen crashes, and at least 1500 people had been killed.

In addition to these crashes, and the China Airlines near-crash, even a cursory inspection of the newspapers shows many more near-misses. In England, for example, two of the most dramatic were in late May 1985, when a 747 overshot the runway in Yorkshire without causing any injuries, and mid-August 1985, when a helicopter carrying the Prime Minister narrowly avoided collision with a jumbo jet at Heathrow Airport. In early April there had also been a catastrophic failure of an RB211 jet engine on an airliner during take-off, though fortunately with less disastrous results than the failure of the Pratt and Whitney of the British Airtours 737; miraculously, no-one was hurt, and through the pilot's skill the plane landed safely.

There seems to be very little linking the six major crashes. The Electra had had a bad reputation in 1959/60, when two early models lost wings as a result of a sympathetic vibration being transmitted from a propellor to the wing, which would then build up into an uncontrollable flutter (a gentle word for such a catastrophic event) and tear itself loose, but, since Lockheed had improved the design, it had become known as a safe, reliable aircraft. The Iberia 727 may have been pilot error, or instrument error, or air traffic control error. The China Airlines 747 was probably a complex interaction of engine failure and conflict thereafter between the autopilot and the human pilot; apparently, even small variations in speed or flight angle can cause a stall or nose dive when the autopilot is in operation, which is the opposite of what a layman (or anyone concerned with air safety) would expect. The crash of the Air India 747 is believed by many to have been caused by a bomb, and the way in which both flight recorders suddenly cut off lends credence to this theory. Delta's Lockheed was apparently caught by 'wind shear', an 80 mph downdraught inside a localised thunderstorm, which thrust the wide-bodied jet down onto the ground, where it bounced once and came to rest a quarter of a mile away, 'like a wall of napalm' according to one eyewitness, after grazing one car on Texas State Highway 114 and completely destroying another, decapitating its driver in the process. The British Airtours disaster was the result of the catastrophic failure of a turbine blade within the jet, which literally exploded through the engine casing and released the furnace heat of the jet itself to ignite the fuel from the tanks which were torn open by the initial explosion. But the JAL 123 disaster stands alone, as the worst single-plane disaster in aviation history: only a collision between two jumbo jets, on a fog-bound runway in Tenerife in 1977, has ever claimed more lives in a single aviation accident.

The JAL flight stands as an almost text-book example of the hazards of flying, a terrifying combination of misfortune, oversight, error, and misjudgement.

To begin with, the seating in the aircraft had been arranged (or 'configured') to jam in as many people as possible. This is increasingly necessary in the cut-throat world of cheap air fares, where costs must be kept to a minimum. There seems to be no question that the jumbo was overloaded, as there was still a massive design safety margin, but it was certainly well loaded: every one of the 509 seats had been sold. The aircraft was also a specially strengthened version of the 747, the 747SR (for 'Short Range'): the discomfort from the overcrowding is obviously less on short runs, but the more frequent take-offs and landings impose an extra stress on the aircraft, which is therefore 'beefed up' structurally. JAL is the only airline to fly this particular 747 variant; they had ten, and now have nine.

Obviously, these super-crowded aircraft are difficult to evacuate — which is one of the reasons for the number of fatalities in the Manchester disaster, though of course that was a 737, not a 747 — and there is a version of the 747 (the 747-300) which is configured for no fewer than *six hundred* passengers, though none of this variant was in service at the time of writing. It might have been possible to evacuate more people, even in an accident like this, if the aircraft were not quite so much like a cattle-truck.

The aircraft in question had also been involved in a rather clumsy landing in 1978, a landing so clumsy that thirty passengers were injured and the tail had been repaired. Examination revealed that where the new tail had been spliced in fewer rivets had been used than were called for in the design specification — in one place, a single row instead of a double one — and a failure stemming from this repair *may* have been what caused the accident, though routine checks should have revealed any flaw. Further evidence supporting this theory is that large sections of the tail-fin, the rudder, and the auxiliary compressed-air unit coaming were found in Sagami Bay, several tens of miles from the scene of the final crash, but it is by no means certain that this was the cause of the crash, and the more one learns about the accident the more puzzling the details become.

The actual time-scale of the disaster was astonishingly short, less than three quarters of an hour from take-off to crash. JAL 123 took off from Tokyo Haneda airport at 6:12 pm, twelve minutes late, and at 6:25 someone on the flight deck reported: 'Immediate, ah, trouble'; English is the international language of aviation. 'Request turn back to Haneda. Descend and maintain 220.' The '220' referred to 22,000 feet, some 2000 feet lower than the height to which the aircraft had already climbed. At 6:27, someone on the flight deck pressed a button which set the '7700' emergency code signal flashing on Tokyo radar screens. Tokyo asked, 'Confirm you are declared emergency. Is that right?' Flight 123 replied 'Yes. Affirmative,' at 6:28 pm.

What happened at that point, and from there on, is mostly conjecture, despite the fact that four survivors — one of them an off-duty JAL flight attendant, 26-year-old Yumi

Ochiai — were able to give eye-witness accounts. Yumi Ochiai reported a bang overhead at the rear of the aircraft, though it was not an explosion. Her ears hurt, but this was due to the loss of pressure; a white mist formed as the cold air of 24,500 feet mixed with the warm, moist air of the cabin. The oxygen masks descended, and a pre-recorded message reminding people how to use them began to play. She felt the aircraft beginning to move like a falling leaf, rocking from side to side; she helped other attendants to show the passengers how to fit a life-belt and how to assume the head-down position in case of a crash.

Meanwhile, still at 6:28 pm, the ground screens at Tokyo Control indicated that the aircraft's flight path was diverging to the right of the flight plan. They radioed 'Fly magnetic 90 degrees', ie due west; at this point, the plane was travelling north-west or west by north-west. The reply was terse but unmistakable: 'But now uncontrol.'

At 6:31 pm, Tokyo Control radioed, 'You are now 72 nautical miles from Nagoya. Do you want to land at Nagoya?' Nagoya was almost twice as close as Tokyo, but for no discoverable reason, JAL 123 opted to return to Haneda. At 6:33, someone on the flight deck reported 'R5 broken.' R5 is Right 5, the rear door through which supplies are normally loaded; it had not been opened at Haneda.

At 6:35, the oxygen ran out. Yumi Ochiai said that she had no difficulty in breathing, but that after the 'falling leaf' motion, the aircraft had gone into a 'Dutch roll', waggling its wings; this may have been the captain trying to steer with the engines, lending more weight to the theory that the disintegration of the tail (or a large part of it) was responsible for the accident. The plane was now heading due north, directly inland, at more than 400 mph — at a right angle to the proper course.

There was astonishingly little communication from the flight deck, although the radio was working. The next that Tokyo Control heard from JAL 123 was the single word, 'Uncontrol', at 6:46; the airport replied, 'Do you want to communicate with Haneda?' Came the reply, 'Yes, please.' At 6:47 JAL 123 asked for the heading into Haneda, again adding, 'Uncontrollable.' Tokyo replied 'Retain magnetic 90 degrees. Can you control?' Once again, there was the single-word answer: 'Uncontrollable'.

Tokyo control could track course of the plane on the radar screens. After the short northern jog, the course was an erratic north-east; at one point, almost due west of Tokyo, the aircraft turned in a complete circle near Mount Fuji before wavering onwards in a north-easterly direction, and then veering south-east before returning to a northerly course. The speed was falling steadily, and the altitude was dropping too: at 6:47 it was at 11,700 feet and 299 mph, a minute later it was at 9850 feet, and a minute later again it was at 7880 feet. It was at 6:49 that Tokyo Control heard someone on the flight deck cry 'Waaah!', the Japanese equivalent of *Aaargh!*, but the aircraft began to *climb* again, reaching 9160 feet. At 6:54, Tokyo Control advised the aircraft that they were 55 miles north-west of Haneda. The acknowledgement — a terse, international 'Roger' — was the last communication heard from the aircraft: when Tokyo Control advised them at 6:55 that both Haneda and Yokota Air Force Base (about 30 miles north-west of Tokyo) were ready for an emergency landing, they had no reply. JAL 123 disappeared off the radar screens at 6:57.

Ochiai had strapped herself into her seat, and knew that they were going to crash. The plane began to drop very sharply; it slammed into Mount Osutaka, in a mountainous part of Japan so difficult to reach that it is sometimes called

'Japan's Tibet'. Fuel from the torn-open tanks flared instantly as the white-hot interior of the engines was exposed in the crash. Ochiai was trapped in her seat by a mass of seats and cushions that had been thrown all around her in the crash; although she did not know it at the time, she had a broken pelvis and broken arms.

The brain produces its own anaesthetics under extreme stress or in extreme pain, and Ochiai slept. She remembers hearing the cries of children before she fell asleep, and hearing a helicopter at one point; she waved, but it was dark, and the aircraft sent over by the Japanese Air Force to investigate the crash could not believe that anyone could have survived the impact, let alone the inferno that followed. In any case, there was no possibility of putting even a helicopter down on the 45° slope over which much of the wreckage was strewn; rescue attempts would have to wait until dawn.

In fact, it was not until 9 am that local firemen reached the site, after a long and difficult climb; paratroopers were lowered by helicopter, though there was no possibility of landing. The aircraft was resting on a narrow ridge, with much of the wreckage (and many of the bodies) spilled into the gullys on either side. Pine trees were scattered like matchwood; small fires were still burning. It looked as if there was no chance of anyone's having survived the crash, but Yumi Ochiai was spotted quite quickly by a fireman, which made everyone redouble their efforts. In a few minutes more, three more people were found alive: Keiko Kawakami, 12, Hiroko Yoshizaki, 34, and her 8-year-old daughter Mikiko. No-one else had survived. The two children were winched into hovering helicopters in the arms of paratroopers; the two women were lifted on stretchers.

Although many of the bodies were horribly mangled,

there were others whose injuries would almost certainly not have been fatal if they had been reached within an hour or two of the crash. The sheer remoteness of the place where the aircraft crashed was ultimately what killed them.

After any accident, the reaction is always, 'If only... If only... If only...' and 'Why...? Why...? Why...?'. In this case, the questions are easy to ask, but almost impossible to answer. Why was the flaw in the tail (if there was a flaw) not discovered? Why did they not try to land at Nagoya? Why did they turn inland, instead of out to sea, where a crash would almost certainly have been at least fractionally less catastrophic? Why that incredible circle near Mount Fuji? Why was there not more communication from the aircraft? Why was the first cry of real fear not heard until 6:49 — and why was that still *five minutes* (twenty or thirty miles) from the crash? And for the 'If only...' reaction: if only they had put the plane down earlier; if only the location had not been so remote....

The more you study the JAL 123 disaster, the more mysterious it becomes. One would expect far more information from the flight deck, even if they were wrestling to keep the plane aloft. Obviously, no-one is going to think clearly in a foreign language when their life is in danger, but it would have been perfectly possible to converse in Japanese. Inexperience on the part of the pilot can be discounted: Captain Masami Takahama was such an experienced pilot that he had been tranferred from internatíonal to domestic routes in 1981 so that he could help with pilot training; he had been flying for JAL since 1966. It may be that it was his skill alone that kept the doomed aircraft aloft for so long.

Ultimately, there are not really very many possible reasons why JAL 123 should have gone down. Pilot error is one, and if a pilot of Masami Takahama's stature could make such a terrible error, we are forced to ask why, and how; but it is not likely. Instrument error is a second possibility, but once again, we can discount it from the evidence available. Mechanical failure, the third possibility, is the most likely — but in an aircraft, a mechanical failure represents a human failure somewhere along the line, either a failure of design (which is unlikely, given the mechanical safety record of all 747 variants) or a failure of inspection and maintenance. Of course, it also means that there has to be something to trigger the fault — in the case of the probable mechanical failure of JAL 123, an aeronautical equivalent of the straw that broke the camel's back. Can we link the comet with this? Well, the theory of the self-fulfilling prophecy is certainly one line — if we expect disaster, we may court it, for example by forgetting safety checks — and the possibilities of direct gravitic, electromagnetic, and solar wind influence on the human brain cannot be discounted. But can we apply the same line of reasoning to the other six major crashes, and to the China Airlines 747 incident?

Beginning with the Galaxy Airlines Lockheed, there was no doubt that it was a very old aeroplane — it entered service in 1959 — but age alone is no great problem with a properly maintained aircraft. More worrying, though, is the fact that the No. 1 engine had already been noticed to be smoky and leaking by an Eastern Airlines pilot the day before the crash; he warned the Electra's pilots, who checked it at the next stop, but found nothing. Two other Electras had also been lost in the preceding nine months, one breaking up in flight over Pennsylvania, and the other crashing during landing in Kansas City: it might just be that the original design problems with the Electra were never completely solved, though the military P-3C Orion variant has proved particularly reliable and successful. However you look at it, though, human error — perhaps in maintenance rather than in piloting — looks likely to have played a part.

The Iberia Airlines 727 crash, into the side of a mountain like JAL 123, appears not to be traceable to any form of mechanical defect, but one cannot rule out either pilot error or air traffic control error.

Current feeling is that the crash of the Air India 747 was due to a bomb, though this ties in with what was said in the last chapter, on terrorism. On the other hand, the inquiry into the crash found no evidence of a bomb (though they did find evidence of explosive decompression), and an open verdict was recorded. One possibility is of course a meteor strike; a piece of interplanetary debris far smaller than the kind of thing discussed in Chapter 9 would be sufficient to destroy an aircraft.

The Delta Airlines Lockheed L-1011 was almost certainly a victim of 'wind shear' — but why should it have happened to a big, reliable aircraft with an experienced pilot? It looks like 'pure' accident, but we have already seen what Jung's theory of synchronicity has to say about 'pure' accident. It is also unfair to imply that the pilot might have been able to do anything about the crash, but he was a good pilot: we have to bear in mind that he *might* have been influenced by the three physical phenomena — gravity, variations in the earth's magnetic field, and the solar wind — even if the likelihood is small. It is also worth noting that the pilot lived for a few minutes after the crash, and that he verbally blamed the air traffic controller, who was completely exonerated by the inquiry. What did the pilot actually see and hear — and what did he think he had seen and heard?

Finally, the British Airtours 737 is the disaster which we

can least readily associate causally with the comet: the catastrophic failure of the fan was something which would not have been readily detectable by routine checks, and the roots of which lie well in the past, whether they were due to a design failure or a manufacturing flaw. Here, we are left only with synchronicity.

One thing which we can say, with some certainty, is that *as far as we know* there is no reason to attribute any of these accidents to *pilot* error, though this is a long way from saying that they are not attributable to *human* error, which includes all manner of maintenance and service staff. All we can say is that there have been some fairly horrific errors in some cases, and equally horrific misfortunes in others.

The record of air crashes in 1985 was bad enough, but to make things worse, there were also several train crashes, including three which definitely qualify for 'disaster' status. Two were in France in August, and the third was in Portugal in September, and in addition to these there were many lesser crashes; in Britain alone there was one on 30th May in Battersea, in London, where 100 were injured but no-one was killed, and another on the Ayr-Glasgow line on 26th September, when 41 were injured but none killed.

The Societé Nationale des Chemins de Fer (SNCF, the French national railway company) actually had three crashes within two months; there was one on 9th July, in which eight died and seventy were injured, another on 4th August, where over 30 people died, and the Argenton-sur-Creuse crash of 31st August killed 43 people and seriously injured many more.

Like the pilot of JAL 123, the driver of the train which left the Gare d'Austerlitz at 9:25 pm on Saturday evening was a very experienced man; he had fourteen years of experience with SNCF, and had been awarded the *medaille d'argent du travaille* — literally, 'the silver medal for work' — in recognition of his service. And yet, at the preliminary enquiry, he admitted that he was travelling at over *three times* the permitted speed.

The usual cruising speed of the SNCF train was 140 kph — about 87 mph. As on roads, though, there are periodical speed restrictions on railway lines; a mile or two before the crash, there had been a 100 kph (62 mph) warning, and he had reduced speed accordingly. A little further on, because of tracklaying work, there was another sign indicating a 30 kph (just under 19 mph) speed limit. Such warnings are not taken lightly: an automatic hooter sounded in the cab, *and the driver acknowledged* it by pressing a switch; if he had not done so, an automatic brake would have slowed the train. Incredibly, though, he continued to travel at 100 kph.

The result was horrific. The train swayed wide as he rounded a bend, and as a result collided with an oncoming mail train. One entire carriage was ripped in two by the impact — and the people who had been asleep in it were ripped apart no less summarily. The firemen and other rescuers who were on the scene in minutes reported seeing blood everywhere, spattered around as if in a slaughterhouse, which is what the train had become. Later, when the immediate task of rescuing the injured and removing the bodies (and parts of bodies) had been completed, one rescue worker told reporters that the impact had been so tremendous that a compartment *two metres* (6½ ft) wide had been compressed into a solid slab of metal 50 cm (just under 20 in) thick.

The driver was charged with 'involuntary manslaughter' and 'involuntary wounding'. At the preliminary enquiry, it was established that fatigue had probably contributed to the accident: he had worked five nights in a row, and at the time of the accident was just one hour and one minute away from the end of his shift. Even so, the scale of his error was incredible. Once again, we must ask: was it the sort of error a man in his position might conceivably be expected to make? The answer is undoubtedly, no. But make it he did — and criminal proceedings were still in process at the time of writing.

The Portuguese crash was also directly attributed to human error. The eastbound Sud Express, carrying Portuguese migrant workers to Paris, collided head-on with a westbound local train from Guarda to Coimbra; it was a straightforward failure of traffic control, apparently due to a poor telephone line between Oporto and Coimbra, with both sides — both experienced men, again — assuming that they had heard each other correctly and that they had got their messages across, which of course neither had done. Over 100 people were killed, and a fireman who was one of the first to reach the scene made the grisly comment that the front coaches were 'burning as hot as a crematorium'; temperatures are estimated to have exceeded 600°C (1100°F), and many of the bodies have simply not been identifiable.

Once again, train crashes are a fact of life, and they happen every year, but we are presented with the extraordinary 'coincidence' that there should be three really serious crashes within as many months. Synchronicity? Or could there really be something which affected the judgement of drivers, technicians, and traffic controllers? Time and again, we come back to the same answer: nobody knows.

8
A NEW PLAGUE?

On 30th September 1985, *Time* magazine published a chillingly precise figure: by that date, they said, 13,228 Americans had been diagnosed as suffering from AIDS, the Acquired Immune Deficiency Syndrome. Of those, 6578 had already died.

Their figures were out of date even as they published them, and they knew it, but they were the best figures available. Within a month of the article's appearance, other sources were estimating that as many as *one million* Americans had been exposed to the disease, which can take two or three years to manifest itself — during which time, those one million people could be infecting others.

The disease is still mercifully rare in Europe, so what follows is based on American figures and the American experience. It is worth adding, though, that the disease has been described by some authorities as 'epidemic' in Central Africa, where it originated, and where it strikes men and women apparently equally, with no distinction between homosexuals and heterosexuals.

AIDS is, to the best of our knowledge, unlike any other disease which has ever afflicted mankind. It was first identified in Central Africa in 1980 or 1981 — there is some dispute — since when it has assumed near-epidemic proportions among homosexuals, with a growing proportion of its victims coming from drug users. Some believe that it is the plague allegedly prophesied in the sixteenth century by Michael of Notre Dame, or Nostradamus, which is scheduled to wipe out four fifths of the world's population by the year 2000, though my own view is that Nostradamus's prophecies always owe more to the reader than to the prophet himself; his allusive style is blindingly obscure, and as a result can be made to fit almost any event — *after* the event.

Before going on to consider the possible cometary connection, which is surprisingly strong if you accept Hoyle and Wickramasinghe's thesis, it is worth looking at just what the disease is, and how it is transmitted.

The body has its own defence mechanisms which reject unwanted invaders. The rejection is a 'learned' response: it takes a while for the body's cells to produce the right *antibodies*, the complex chemicals which block the reproduction of the invaders. Long before inoculations were ever invented, it was known that there were many diseases which, if the sufferer survived the first time, would never come back or which would at worst only return in a very much milder form. The modern practice of inoculation derives from a reversal of the same process; we deliberately infect ourselves with a weakened version of the disease, usually in the form of 'killed' pathogens, in order to 'teach' our cells how to make the antibodies. Then, if we are exposed to the disease in its full-strength form, our bodies can mobilise their defences very much faster — so fast, in fact, that the disease does not have a chance to establish itself.

AIDS makes a mockery of all this, by striking at the very mechanism which allows us to produce antibodies. The normal sequence of events, in the absence of AIDS, is for a *macrophage* — a type of white blood cell — to 'recognise' the invader, and then to pass the message on via *T cells* (another kind of white blood cell) to *B cells*, yet a third type, which actually produce the antibodies. The AIDS virus strikes at the T cells, where in the familiar viral pattern it takes over the cell and forces it to produce more AIDS virus. The B cells are, therefore, never 'alerted', and AIDS is free to spread throughout the body.

There is more to it than this, though. Because there are now no new antibodies being produced (hence 'Immune Deficiency Syndrome') the body is vulnerable to attack from literally anything else that happens to be about — viral, bacterial, fungal, or cancerous — and *this* is what usually proves fatal. AIDS can, therefore, take a large number of forms, most of them extremely unpleasant; even diseases which are very poorly adapted to man, and are normally found only in animals, can take a hold, work their way through the body, and kill the victim. If he or she is lucky, the progress of the killer disease will be rapid; if he or she is unlucky, it can be a very lingering death indeed, with ulcers, skin cancers, fungus infections of the skin and digestive tract, pneumonia, and even the common cold all adding up to a package which does not kill the patient cleanly, but saps his vitality day by day, a sort of death by attrition.

At the time of writing, there seemed to be no certain leads on a cure for AIDS, but there are two modest causes for consolation. The first is that not everyone who is exposed to

AIDS, or even infected with it, necessarily shows the symptoms; their immune system comes to an equilibrium with the virus, and they do not suffer from any of its horrifying effects, though some do display what is known as ARC, or AIDS-related complex, which shows itself as general debility and aches and pains, without necessarily developing into AIDS. A frightening corollary to this, though, is that these people can remain AIDS carriers, perhaps even unwittingly, and thereby infect other people even if they do not suffer themselves. Paradoxically, however, even AIDS is not itself immune to an antibody reaction, and it is this which gives researchers considerable hope in their hunt for a cure.

The second consolation is that it is *at least in the Western world* quite a difficult disease to catch. First christened the 'Gay Plague', because of the predominance of male homosexuals among its victims (up to three quarters, even now), the only proven transmittters of AIDS virus are semen and blood; it has been detected in spittle and urine, but there is yet to be a proven case of transmission by either medium, so even passionate kissing is not *so far as we know* likely to transmit the disease — and given the intensity with which the subject is being studied, it seems likely that our knowledge is correct.

The reason why it strikes so hard among the male homosexual population is simple, but unpleasant to consider. During anal intercourse, the wall of the colon is often torn or injured. This enables the blood of the passive partner to mingle with the semen of the active partner, which is how the disease is transmitted from the active to the passive partner (transmission in the opposite direction is much less well documented). Of course, the same applies in male/female anal intercourse, and it is believed that some women sufferers may have acquired the disease in this way. It is also possible that swallowing semen can pass the disease on, so heterosexual and homosexual fellatio may also play a part.

The other major group of sufferers, intravenous drug users, pass the disease from one to another with dirty needles: even the minutest quantity of blood from an AIDS victim is enough to infect the next user of the needle, provided the use is fairly immediate: the AIDS virus cannot survive for long outside fresh blood, which is why the fears about toilet seats, etc, are unfounded. About a fifth as many men contract AIDS this way as from homosexuality, but over half the women who have the disease are believed to have acquired it in this way.

After these two major groups, there are three minor groups. First, there are those who have acquired the disease from heterosexual intercourse; so far, there are very few documented cases of this, and the transmission has been male-to-female by a factor of about ten to one. Secondly, there are those who acquire AIDS either in the womb or during birth; the mechanism is unclear, and there are mercifully few documented cases, though they may be expected to increase. Finally, there are those who have acquired the disease via transfusions of blood donated by AIDS sufferers; this accounts for about 1 per cent of American men and 10 per cent of American women who suffer from the disease. Even those who try their best not to make value judgements based on conventional morality find these cases the hardest to accept: any one of us could be involved in an automobile accident and be infected by a routinely given blood transfusion, though haemophiliacs (who often depend on frequent blood transfusions just to survive the risks of day-to-day life) were particularly at risk. Note the use of the past tense here: donated blood in the US, the UK, and most other countries is now routinely tested for AIDS, so the risk from this particular source is very slight and decreasing.

As already mentioned, the epidemiology of the disease is rather different in Third World countries, especially among black victims. There could be several reasons for this. One is that it is a separate strain of the disease, which may (or may not) spread to the West. Another is that there are racial variations in susceptibility; a well-known example of this already exists in the form of sickle-cell anaemia, which although a disadvantage in itself confers increased resistance to malaria, which makes it (on balance) a survival characteristic. A third is that the disease has actually been around in Central Africa for longer than we think, and that it *is* actually much easier to catch than we think; it just takes longer to manifest itself in a more heterosexual society in which people do not stick dirty needles into their arms. Yet a fourth is that there may be dietary considerations which predispose potential victims to AIDS, and a fifth is that other diseases may potentiate the AIDS virus so that it can strike more easily: herpes has been the subject of considerable study on this front.

What, though, might the link be between AIDS and comets? So far, throughout the book, I have referred to Hoyle and Wickramasinghe's theory without explaining it in any detail. In essence, it suggests that Darwinism is neither a necessary nor a sufficient mechanism to explain the diversity of life on Earth. It points to an undeniable lack of fossil evidence, and with the aid of mathematical arguments on probability which I have yet to see refuted, it shows that the likelihood of life's evolving on Earth is negligible in comparison with the likelihood of its having evolved on comets. Hoyle and Wickramasinghe (hereafter referred to as H&W in the interests of brevity, but with no disrespect intended) argue with considerable force that atmospheric and temperature conditions on the early Earth, to say nothing of the mixture of chemicals available as the basic 'building blocks' of life, were so unlikely to have been able to engender life that the possibility is not really worth considering, while a good-sized 'dirty snowball' could retain warm pockets within, either stored solar energy from a close stellar fly-by or as a result of nuclear reactions, which *would* provide ideal conditions. These pockets would eventually freeze as the snowball cooled, until they reached absolute zero, but another pass around the sun could thaw them out and bring them into the comet's envelope, from where they could be spread by the comet itself *or by the pressure of the solar wind acting upon a comet's tail.* There is no question about the ability of viruses to survive this sort of mistreatment, and more complex organisms (including bacteria) can almost certainly do so.

H&W's thesis is supported by a great deal of argument, much of it anecdotal but nevertheless supported by references, which answers all the main points which 'everyone knows' rule the possibility of an extraterrestrial origin for life of any kind, including the various forms of 'hard' stellar radiation (X-ray and ultra-violet), the way in which asteroids usually burn up as they enter the earth's atmosphere, and so forth. The biggest hurdle, the question *why* extraterrestrial life should have any common ground with terrestrial, they answer with breathtaking simplicity: there are very few possible patterns for life, and because of the incredible replication or reproduction rate of biological entities (especially viruses) a basic 'building block' appearing early in the history of the universe — well before the formation of the Earth — could 'scoop the pool' and become (literally) universal. A particularly interesting point that they make concerns the resistance of bacteria and viruses (and insects, though to a lesser extent) to hard radiation, extremes of temperature, and very high and very low pressures which do not exist naturally on earth: this resistance constitutes a magnificent *anti-*Darwinian argument, as there is obviously no Darwinian advantage in being able to survive hazards to which the organism would never be exposed — but it makes absolute sense when considering possible interstellar journeys....

I must confess that I had not read their books until the idea for the present book came to me, and that I initially approached them with a scepticism bordering on scorn, but, by the time I had finished them, I had to admit that their theory fitted the facts *at least as well* as any other theory, and better than most. There would be no point in trying to summarise H&W's three books here, but AIDS seems to be a prime candidate for their argument: a disease which appears from nowhere, against which we have no defences, and which struck (at least initially) in a curiously random pattern.

On the other hand, even if one accepts their thesis in full, there is absolutely no need to tie AIDS in with Halley's Comet. H&W make the point that so-called 'empty' space may be very empty by our standards, but it is still far from 'empty' in absolute terms. They calculate that the possible number of bacterial, viral, or DNA/RNA containing particles which strike the Earth each day could be as high as 10^{20}, or 100,000,000,000,000,000,000 (a hundred billion billion), of which only a few need to get through in order to seed the earth with disease. An intriguing addition to the theory, though, is that not all 'diseases' are bad for us; indeed, it is central to H&W'S theory that most if not all of the significant advances in evolution have come as a direct result of comet-borne 'disease'.

It is possible therefore that Halley's Comet and other comets are linked with disease, and the present AIDS 'plague', at least if you accept H&W's argument. I think I do — but you will have to decide for yourselves. To help, H&W's three books are *Lifecloud*, *Diseases from Space*, and *Evolution from Space*, and the authors are Sir Fred Hoyle, FRS, and Professor Chandra Wickramasinghe.

9
THE ULTIMATE CATASTROPHE

There is no doubt that, in the past, the Earth has been struck by good-sized meteorites or cometary particles. It is virtually certain that, in geological time, it will happen again. The only question is when.

The good news is that Halley's Comet is *not* going to strike the Earth — not on this pass, or on any pass in the future whose orbit has yet been calculated. There is the extremely remote possibility that if Halley's Comet broke up at *precisely* the wrong time, a fragment *might* hit Earth, but it is not worth worrying about; it is about on the same level of probability as an invasion of giant spiders from Mars — at least on this pass.

The bad news is that we do not have the faintest idea when the next major meteorite strike will be. The popular scenario, which has occupied so many 'penny dreadfuls', of the astronomers sighting a full-blown comet which they calculate will strike the earth in X months time, thereby throwing the whole world into panic, is a lot less likely than a far more frightening scenario in which a relatively small, dark chunk of cosmic debris which the astronomers never even notice strikes utterly without warning, as apparently happened in Siberia on 30th June 1908.

The Stony Tunguska River is in an impossibly remote part of Siberia. The nearest town to the site of the Tunguska Event (or Tunguska Incident — both terms are used) is 40 miles away. The inhabitants saw a fireball sometimes reported to have risen to a height of twelve miles; felt a wave of intense heat; and were then subjected to a shock wave which cracked windows, threw objects from shelves, and indeed knocked several people over. Nomads slightly nearer the putative blast site had their tents knocked over, and the passengers of the Trans-Siberia Express, about four hundred miles away at the time, saw a blue fireball hurtling through the sky and heard the explosion; the bang was sufficiently loud that the engine driver thought that he had burst a boiler tube, and stopped the train.

Because the area was so remote, and because the Russian revolution intervened, it was almost two decades before an investigative party arrived. They found hundreds of square miles of forest that had been flattened by the blast; the trees radiated outwards from 'ground zero', the centre of the explosion, and were scorched and blasted by immense heat. Much has been made of the fact that neither they nor subsequent expeditions ever found any decent-sized fragments, or even a large crater, just a number of small craters and meteoric grains 0.1 mm ($1/250$ inch) in diameter; on this evidence, many have argued that it could not possibly have been a meteorite (which would have left a crater and more traces) and have suggested all manner of fanciful explanations, including an exploding spaceship; an asteroid made of antimatter; a natural nuclear bomb (!); or a miniature black hole which bored straight through the Earth and emerged through the Atlantic Ocean — a sort of reverse China Syndrome!

In fact, we already know of one celestial body which would fit the facts very well, and which does far less violence either to the known facts or to common sense: the 'dirty snowball'. If the Tunguska meteorite was actually a part of Encke's Comet (and it coincided with a daylight meteor shower believed to have derived from a partial break-up of Encke's Comet), then it would have consisted mostly of ices of various sorts (water, ammonia, methane, etc) and dust. The ices would evaporate explosively, leaving no trace, and the dust particles would be of about the right size to account for those found.

If it was a 'dirty snowball', it may have been a big one by Earthly standards, but it was minute on a cosmic scale: about 150 feet (40 metres) in diameter, and weighing about 50,000 tons. Such an object would never be spotted by a telescope, and would only be visible to the naked eye a few minutes before impact. Because it could be travelling at anything up to 100,000 mph when it hit the Earth's atmosphere, and would still be travelling at several thousand mph when it burst, no-one who was in the impact zone would ever even see it, and no-one who saw it would be close enough to the impact zone to be harmed.

The Tunguska meteorite was about as powerful as a fair-sized H-bomb; an Israeli scientist (Ari Ben Menahem) calculated that a 12.5 megaton bomb bursting about five miles up would have been sufficient to cause the same levels of damage as were reported at Tunguska, although earlier estimates were rather higher — anything up to 100 megatons. Such variations are perhaps inevitable in the light of the unknown factors involved, but they do bring us to the terrifying possibility that a modern Tunguska Event *would be taken* for a nuclear attack, especially if it hit a major city in the United States or the USSR, and that it would therefore provoke a nuclear reaction. The men who build and test nuclear weapons on both sides are aware of that possibility; whether the men who have access to the button that launches them are as well informed is another question.

In any case, there is another possibility which makes even nuclear war look like pretty small beer; after all, it is likely that, in the remotest areas, man would still survive a nuclear war, and that after a few thousand years, civilisation would be re-established. There are, of course, legends that this has happened....

The really alarming possibility is a strike by a major piece of cosmic debris — a full-blown comet, an asteroid, or an *apollo*. An apollo is a 'worn-out' comet; we have already seen how comets lose a part of their mass as they pass the sun, and across a few million years all the ices will be boiled off, and what is left is a lump of dust, perhaps bound together with tarry organic chemicals, travelling at cometary speeds. It has been calculated that such objects strike the Earth perhaps four times every million years, and that, when they do so, the explosive force of a body (say) a thousand yards in diameter is equivalent to that of a *one hundred thousand megaton* hydrogen bomb — in other words, more than the entire stock of nuclear bombs which has been exploded in human history, or a thousand times greater than the total energy of the great Krakatoa volcanic eruption of 1883, when *four cubic miles* of rock were hurled into the air and the explosion was clearly heard *two thousand miles away* in Australia.

Although we have no direct evidence of meteor strikes on

this scale, we do have indirect evidence, to which I shall return shortly. We do, however, have direct evidence of a major nickel/iron meteor strike in Meteor Crater, Arizona, where (according to the most widely accepted estimates) a ball of nickel-iron some 80 feet in diameter and weighing 63,000 tons hit the Earth at about 36,000 mph (60,000 kph) and buried itself a quarter of a mile deep. One of the most alarming things about the Arizona meteorite is that the most distant estimate of when it struck places it only 50,000 years in the past, and that some place it as recently as 5000 years ago — or very nearly in historical (as opposed to prehistoric) times. It is worth noting, too, that the legend of a great flood is common to almost all cultures, except Tibet (average altitude 13,000 feet/4000 metres). Tibet has a legend of a shoreline . . . .

Meteor Crater, Arizona, is about 1300 yards (1200 metres) across. Chubb Crater, in Quebec Province in Canada, is 2 miles (over 3 km) across, and the Vredefort Ring, near Johannesburg in South Africa, is 30 miles across, though the latter two *may* not be due to meteor impact; there is a case for their being very old volcanic formations. Even so, the Arizona meteor was pretty small on a cosmic scale, and impacts of similar sized objects can be expected about every 100,000 years. This does not mean that they will come, regular as clockwork, every 100,000 years; it means that a million years may pass without a single impact, and that ten impacts might then occur within a mere 50,000 years. According to some authorities, the city of Pittsburgh was nearly destroyed by a meteorite — again, probably a 'dirty snowball' — on 24th June 1938, when an air-burst near miss led some people in that city to imagine that the West Winfield powder magazine nearby had exploded. You may be seconds from destruction now . . .

Or you may be 13,000,000 years from it, which means that it is not a matter of immediately pressing concern, at least in this incarnation. While astronomers agree that cometary strikes (or strikes involving a part of a comet, as at Tunguska) and meteor strikes are a certainty, opinions differ widely about their frequency. One of the most 'respectable' theories currently being aired is that there is a 26,000,000 year cycle of planetary bombardment by meteors, which is currently about half over.

The theory was triggered by Alvarez *père et fils*. Walter Alvarez, a geologist of international repute who was working at Gubbio in Italy at the time, was struck by the presence of a thin sedimentary layer between Cretaceous deposits — rich in fossils, and dating from the time of the dinosaurs — and Tertiary deposits, in which fossil records are very rare indeed. He sent samples of the sedimentary interlayer to his father, Luis Alvarez, a Nobel Prize-winning physicist in California, who found iridium levels *thirty times* as high as are usual in the Earth's crust.

Iridium is a rare element not unlike platinum, a hard silvery metal. It occurs deep in the Earth's core (perhaps 3000 km/2000 miles down), in asteroids and comets, and in some kinds of interstellar dust. Realising that it could almost certainly not have come up from the core, especially in a way that could form a sedimentary rock, and knowing that interstellar dust is certainly far too thin to leave any traces at all, they suggested that the iridium was of asteroidal or cometary origin: in short, that a major comet strike about 13,000,000 years ago might have been what caused the extinction of the dinosaurs. The idea of a 'nuclear winter', in which the dust clouds caused by massive nuclear bombing would cause a prolonged lowering of the Earth's temperature, is by now familiar to most people: the Alvarez hypothesis postulates a massive comet strike followed by a 'cometary winter', which would be enough to wipe out whole lines of evolution; their theories preceded the nuclear ones, so the concept of 'nuclear winter' may be said to be based on their theories.

Since the Alvarez's original work in 1977/78, many other scientists have also become interested in the idea of 'cometary winters' as catastrophes of sufficient magnitude to explain the discontinuities in the fossil record; indeed, this provides an excellent alternative to Hoyle and Wickramasinghe's thesis as a means of providing the sharp 'kicks' which evolution appears to have experienced from time to time. As they searched the fossil record, a fair degree of periodicity became apparent: every 26,000,000 years or so, there did seem to be a major extinction of several lines of evolution. Why?

There are two current theories. The more dramatic postulates a 'death star', christened Nemesis after the Greek goddess of vengeance or retribution, particularly the kind of retribution which overtakes the proud and haughty. Nemesis orbits around our sun on a 26,000,000 year cycle, and as it passes through the Oort cloud of comets on the outskirts of the solar system, it dislodges many — perhaps thousands — and hurls them among the solar system like gigantic thunderbolts. Every 26,000,000 years, the Earth must face one of these showers; the next is due 13,000,000 years from now.

The other theory is merely a slightly less dramatic variation of the same thing. It postulates a 'Planet X' (straight from the pages of *Amazing Science Fiction!*) with a much smaller and less elliptical orbit, comparable with that of the outer planets *but not in the ecliptic plane* in which the other planets move. Inevitably, the orbit of 'Planet X' is subject to considerable gravitational influence from the other planets, and is constantly changing, but, like Nemesis, it intersects with the Oort cloud every 28,000,000 years or so — the figures do not agree precisely, but, equally, the two million year variation is well within the experimental, theoretical, and geological range of the available data.

There is little to worry about from either Nemesis or Planet X in the immediate future. If man still exists in 13,000,000 years' time, then we may hope that he will have solved the secret of interstellar travel, and that he will be able to evacuate the Earth, or that, alternatively, he may be able either to divert the path of Nemesis/Planet X, or to protect his planet from meteor strikes in some other way. There are some pretty far-out theories between these covers; this particular one, though, is being taken sufficiently seriously that astronomers are currently searching the heavens for a dark body which might be either Nemesis or Planet X.

The possibility of a major meteorite or cometary strike *not* related to Nemesis or Planet X is another matter. If we can avoid panic, even a meteorite or cometary fragment strike on the scale of Tunguska or the Arizona Crater will not be the final disaster, though it could change the face of the world as we know it. We could lose millions of people in the initial strike, both from impact and *tsunamis* (tidal waves), to say nothing of the likelihood that the impact would very probably trigger off several major earthquakes; millions more could die of famine afterwards. It is a frightening prospect — and it could happen at literally any moment.

10
APPROACHING PERIHELION

The Chinese have an ancient curse which says, 'May you live in interesting times.' We live in interesting times.

This book was completed only a few days after the first photograph of the 1985/86 apparition of Halley's Comet had been published. I delayed giving the manuscript to the publishers for as long as possible, because every day there was something new; the hijacking of the *Achille Lauro*, the mud-slide in Puerto Rico which killed 500 people, the earthquake in central Russia, of which details were still emerging when I could delay no longer. It certainly looked as if the procession of disasters which characterised 1985 was continuing unabated.

It takes a man more rash than I to write a book of confident prediction; essentially, this book is historical. But in this last chapter it is worth speculating rather more wildly than one would normally do, just in order to see whether any of the possibly comet-associated phenomena described in the foregoing chapters actually amounts to anything. It is also worth looking at what — if anything — we can do as individuals and as groups to avert *comet catastrophe*, the very title of this book.

In the second chapter, I canvassed four separate lines of inquiry, and throughout the book I have tried to apply these lines to the various 'disasters' described: I put the word 'disasters' in inverted commas because, viewed on a cosmic scale, they are really fairly trivial — what are a few hundred or even a few thousand deaths on one small planet in one solar system, when viewed against the vastness of the cosmos?

Perhaps, though, the very human scale of these disasters is exactly what we should expect. All through the book, the extreme weakness (on a cosmic scale) of all the forces involved has been stressed; is it not likely that these cosmically trivial forces might be expected to result in equally cosmically insignificant events, even though these events when viewed from our human perspective are anything but insignificant? Is it not also likely that *if* there is any connection between disaster and Halley's Comet (and its immediate predecessors West, Bennett, Kohoutek and Tago-Sato-Kosaka), then the scale of disaster should increase as the comet draws closer? Bearing in mind what has already been said about inverse square laws, it is important to remember that when the comet is *twice* as close, its effects will be *four times* greater; and when it is *four times* closer, the effects will be *sixteen times* as great.

The first of our lines of inquiry, that of *synchronicity*, may be viewed in two lights. On the one hand, the synchronicity of the comet with other events is well worth watching; will it presage any or all of the traditional disasters such as the fall of kings, the destruction of dynasties, war and famine? Now that there are far fewer princes in the world, should we extend our definition of 'great men' (and great women, of course) to include politicians? The cynic, looking at the present incumbents of most offices, will surely answer 'No!', but it is worth remembering that Rome was for much of its history a republic, and that the astrologers then were as much concerned then as they were at Imperial periods. The trouble with this line of inquiry is that few of us feel able to do anything to influence the way of the world, and that the whole of the twentieth century has been such a political tumult that it would be hard to distinguish anything out of the ordinary, even if political disasters occurred right, left, and centre.

A much more interesting approach is to view synchronicity in our own lives. If a (relatively) highly visible comet such as Halley's does betoken unusual times, we can conduct a simple experiment. Instead of writing things off as 'coincidence' or 'trivial' or simply 'mistaken', as we so often do, we can try to see whether there is actually a higher degree of interrelatedness between what we normally regard as different aspects of our daily life. Keeping a diary for the next six months, and noting all coincidences, 'predictive' dreams by ourselves and our friends, and so forth, could be a fascinating exercise. Of course, in order for this to be a genuine experiment, we should also keep a similar diary in (say) a year's time, when the influence of Halley's Comet may be expected to have diminished dramatically — unless, of course, a new comet is discovered before then!

Any such experiment must, of necessity, be passive: there is not much we can actually *do* in order to affect the course of events. Moving on to the next line of inquiry, which deals with known physical phenomena, we can act (to a limited extent) as well as observing.

From the gravitic point of view, there is absolutely nothing which we can do in order to avert any disasters which can happen - but we can act to minimise their effect on ourselves. The most obvious thing is to stay out of possible danger zones, which means, basically, out of areas of known earthquake or volcanic activity, or which could be subject to *tsunamis* (tidal waves). It is generally agreed that the San Andreas fault, in particular, is overdue for catastrophic release: geologists have calculated on the basis of past tremors that what Californians call 'the big one' will be much greater in magnitude than the San Francisco earthquake in the early part of this century, and it is worth reflecting upon the point that Los Angeles is a very much bigger town now than it was three quarters of a century ago, so the destruction there (even with modern 'earthquake-proof' building techniques and materials) is likely to prove far greater than it was in San Francisco in 1908. Nor is San Andreas' well-known fault the only thing which the inhabitants of the Pacific coast of the United States have to worry about: the possibilities of *tsunamis* along the immediate coastal strip are far from negligible.

It might seem that as long as we avoid California, Japan, the Himalayas, and other places notorious for earthquakes, we are fairly safe; but it is worth remembering that even dear old geologically stable England has had earthquakes in the past, and that in 1985 there were earthquakes in Assam (January, 27 killed), Mendoza, Argentina (February; an estimated 40,000 homeless, number of dead unknown); Santiago, Chile (March, over 140 killed); and non-fatal 'quakes in the Philippines, southern Spain, and even Denmark and southern Sweden! It is also worth pointing out that virtually no attention has been paid in most of Europe (except perhaps in Iceland and parts of Greece and Italy) to earthquake-proofing new buildings. If you work in a high-rise building, perihelion might be a good time to take an early vacation, and I for one shall not be booked into a high-rise hotel in February! As omens of earthquakes, pay particularly close attention to the behaviour of wild and domestic animals: there are several well-documented cases of (for example) ducks and swans leaving a lake *en masse* a few hours before an earthquake, and many people have reported that their dogs had been behaving strangely, whining to be let out, before 'quakes.

As an aside here, it is worth remarking on the strange indifference of Californians to the fact that they are sitting on a geological time bomb. It is something which always fascinated me, particularly when I lived in southern California for a while; there is something of a suspension of disbelief, a feeling that 'It can't happen to me', when you are there, which cannot be explained in wholly rational terms. Certainly, minor earthquakes are nothing like as frightening as you anticipate, and once you have experienced one or two, you stop worrying about the small ones; but why you should stop worrying about the big ones as well is something I cannot explain, and I have lived there and even driven through the town of San Andreas.

No matter where you are, or where you live, it is worth laying in a stock of convenient canned or other long-life (non-refrigerated) foods *which can be eaten cold*, as well as a few bottles of Perrier water (or even tap water!); disruption of power lines and water mains is likely even in a small earthquake, and, while such modest preparations are unlikely to make a difference between life and death, they can make the difference between moderate and major inconvenience. More drastic 'survivalist' type preparations are hardly worth the effort, though I must admit that if I am in the Western United States I shall take good care to be sure that my car is well filled with gas (and that I have a couple of jerry-cans spare); that I also have a couple of jerry-cans full of water; and (though this is a touch melodramatic, and probably says more about me than about any probable danger) that my guns are loaded and to hand.

The same is true of electromagnetic phenomena as of gravitic: all you can do is to try and schedule your time so that you do not rely too heavily on radio communication during the few weeks surrounding perihelion. This may not sound difficult, but if you make any international telephone calls, the chances are that your calls are routed via satellites, which are peculiarly prone to interference; it is not a period in which to try to transact international business, for this very reason.

It would be a good idea, too, to monitor the level of air and train crashes: it may well be that the spate of crashes which characterised 1985 was unrelated to any form of cometary influence, whether purely physical (in the form of disruption of communications) or in some less well-understood way involving human physiology, but if the pattern of air disasters in particular remains constant (or even increases, which is what we might expect as a result of the action of an inverse square law) it would be a pretty potent argument for avoiding air travel!

When it comes to the solar wind, it is worth bearing in mind the increased possibility of 'soft failure' on computers. It seems likely that cosmic rays can occasionally affect information stored on computer tape (or other magnetic media), by direct interaction with the molecules in whose magnetic fields the information is stored. The resultant computerised snafu (confusion) is referred to as a 'soft failure', because it is neither a hardware failure (in the accepted sense) nor a programming error, though it is still effectively a software failure. The possibility of 'soft failure' may increase significantly at around the time of perihelion, so be prepared for demands for millions from the utility companies and the tax authorities, courtesy of their computers! If you use computers yourself, make security back-up copies of all programs and data; the chances of a 'soft failure' are remote enough anyway, but when you are dealing with two copies, they become utterly negligible.

If gravitation, electromagnetism, and the solar wind are purely physical phenomena which have no *direct* impact upon the human frame or mind, then the above precautions should be fairly adequate; at least, they are all that is reasonably practicable. If, however, there are direct effects then there is a good case for being very circumspect in your own affairs. In business, for example, it might be as well not to make any important or far-reaching decisions during the few weeks either side of perihelion, and the same would apply to personal decisions such as marriage, divorce, etc. On the other hand, there is the distinct possibility that, if the phenomena mentioned above actually do affect people, they may affect some for the *better* rather than for the worse. The difficulty here obviously lies in trying to be objective about it; as anyone who has ever had one drink too many will know only too well, it is as easy to overestimate your own competence and ability as to underestimate it, so seek the advice of others before trying to put any of your brainstorms into effect (but remember that they too may be euphoric or depressed . . .).

The third line of investigation, about 'diseases from space', is probably the one about which you can do most — and even if the whole theory turns out to be pure rubbish, it will do you some good anyway. All you need to do is to try to build up your general level of health, on the simple basis that if there are diseases about (whether comet-borne or not) you are less likely to catch them .

A healthy, balanced diet is an obvious start, and if you believe in vitamin or other supplements then take them too; unless you take them in insanely large doses, the very worst that can be said of them is that they will lighten your pocket. Trying to keep your weight down is a good idea, and so is giving up smoking; on the other hand, 'trick' diets and quack medicines (such as the 'anti-comet pills' which were sold for the 1910 apparition) are likely to do you no good at all, and can even increase your susceptibility to disease by upsetting the general expectations of your body; too-abrupt changes in eating habits, lifestyle, etc, can actually do more harm in the short term than sticking to familiar patterns, even though these may be known to be less healthy.

Exercise is slightly more problematical. As an aid to general fitness, its value is indisputable; the only problem comes when you realise that, in order to take exercise, you have to venture out of doors, which is precisely where we might expect to find these biological unpleasantries lurking, and the deeper inhalation associated with exercise might also be expected to introduce more airborne pathogens into the lungs. On balance, the increased fitness due to exercise is more likely to ward off disease than the increased exercise is to lead to your catching any form of infection — whether comet-borne or otherwise — and exercise is therefore to be recommended, though the points about drastic changes in lifestyle which have already been made should also be borne in mind.

Our last line of investigation, the self-fulfilling prophecy, is also a comparatively easy one in which to defend yourself — at least against your own actions, if not against those of other people — though there is a 'catch-22' built into the situation. By determinedly ignoring the possibility of catastrophe, you are (inevitably) focusing your attention upon catastrophe, and the result is like the childhood challenge which exhorts you *not* to think about (say) a white horse for one minute: because you know that you are supposed *not* to be thinking about the horse, it keeps forcing its way into your mind, even though you might normally never think a single thought about white horses from one month's end to the next.

Obsession is relatively quickly disposed of: medical treatment, or counselling from a wise friend, minister, or even professional counsellor is the answer if you cannot break out of it on your own.

Distraction is more of a poser, for the very catch-22 reason referred to above, but the best advice here is simply to get along with your everyday life; if you concentrate upon what you are doing, rather than upon what you are not doing, you stand a very much better chance of completing your task successfully! This applies whether the 'task' in question is work of some kind, or simply recreation: if you are concentrating on the bought ledger, or on hang gliding, rather than on Halley's Comet, your chances of personal disaster are markedly reduced!

Several religions teach techniques for avoiding distraction, though it is important to distinguish between those which are based on 'no-thought', such as Buddhism, and those which are essentially based on a counter-irritant, such as a fear of damnation or a hope of heaven. By appreciating *shunyata*, voidness, as the paradoxical centre of being, you can look at all phenomena with equanimity: if instead you are concerned with external forces over which you have no control, it is much *more* difficult to stop worrying.

In fact, there is room for considerable speculation on the relationship between religion, catastrophe, and the comet. On the one hand, the authoritarian religions (Judaism, Christianity, and Islam, all of which are more closely related than their adherents like to admit) have become detectably *more* rigid. In Israel itself, organisations such as the Gush Emunim are in the ascendant. In the Catholic church, Pope John-Paul II has reproved the Jesuits because they have repeatedly subjected dogma to intellectual criticism (a problem which Popes past and present have fought) as well as taking a hard line on such issues as birth control and laicization (leaving the ministry). Among Protestants, there has been a surge in the popularity of fundamentalism; the Southern Baptists are flourishing, and in their 1985 assembly reaffirmed by popular vote their belief that every word of the Bible is literally true. The influence of Islam, the third of the near Eastern religions, is higher than it has been for years, or decades; the rise of Islamic fundamentalism has been visible everywhere. On the other hand, the more inclusive religions, especially Buddhism, have also been gaining converts at record speed: Buddhism has probably been expanding faster than at any time for a thousand years, perhaps longer, as an intellectual and rational framework for existence which appeals to those who cannot accept authoritarian and dogmatic religion.

Either way, people seem to need a moral and ethical framework more than at any time in the recent past, but the alarming part is that within the Christian/Islamic tradition there has always been a significant group of movements who preach 'the Faith or the Sword'; with the rise of Christian fundamentalism in the United States (which is already armed to the teeth with nuclear weapons) and the possibility that states like Libya or Iran could use oil revenues to buy nuclear weapons, via terrorism if necessary, religion may be as much a destabilising influence as a stabilising one.

The same arguments as were used above can also be used when considering coincidence, the third division of our analysis of self-fulfilling prophecy. Earlier in this chapter, I recommended trying to keep a deliberate diary of coincidences, but the time has come to add an important rider: do so *without worrying about them.* The important thing is to accept them for what they are — to wit, simply a part of the seamless fabric of life — and to accept that there are times when you can act, and times when you cannot. If you act when it is time to act, and refrain from action when action is useless, your whole life will be very much easier, and this is a message which is taught alike by Jehovah, Lao Tsu, the Buddha, Christ, the Prophet, and all enlightened teachers.

So: what conclusions can we reach in our consideration of the relationship between comets and catastrophes? It seems to me that there are four possibilities.

First, there may be absolutely no causal or acausal linkage between the two — in which case, we have nothing to worry about, except perhaps just why there have been so many disasters recently: if it is not the comet's fault, whose fault is it?

Second, there may be a causal connection between the two, a causal connection which could be the result of a number of different phenomena, some known, some imperfectly known, some unknown. For example, aircraft instrument failure could be linked with electromagnetic and solar wind disturbances, and it is not impossible that earthquakes are linked with gravitational changes, though the latter is much less likely, as the mass of Halley's Comet is no more than that of a good-sized mountain, which is negligible in comparison with the Earth's gravitational field. There may also be something in Hoyle and Wickramasinghe's theories about comet-borne diseases. The *unknown* phenomena, particularly those affecting human behaviour, may be more important than those which are understood; for example, it is only very recently that human mood has been linked to the concentration of positive and negative ions in the atmosphere (which we may reasonably expect to vary detectably under certain conditions which could happen during a cometary fly-by) and it may be that the 'solar wind' effects are much greater than we think. We can keep the effect of this upon ourselves to a minimum, and in some cases (for example, by not flying) we can stop others affecting us; a more alarming case would be if politicians or generals fell under some sort of influence, as described below.

Third, there is the strong possibility that a self-fulfilling prophecy effect exists — an effect which we can keep to a minimum in its impact upon our own behaviour, but which we cannot so readily answer for when it comes to others. There is a good case here for prayer that those who are badly affected will not be in a position to affect us, because if a single group of politicians or the military decided to act out some bizarre Doctor Strangelove fantasy of destroying the world their opponents would have no alternative but to react in accordance with current military thinking — which means the destruction of the world as we know it. It is all very well to say that politicians and generals are not that crazy, but does their everyday behaviour actually bear out that argument?

Fourth, there may be an *acausal* connection between the comet and catastrophe, along the lines described in Jung's theory of synchronicity: in other words, certain things may happen at the same time, or very close together in time, in a way which is not logically or causally related, but which still indicates an inextricable interlinking of the two (or more) events.

The various nuclear nightmare scenarios are real enough, but (in a sense) carry their own solutions with them, in that few (if any) of us will live to worry about them. Likewise, although the possibility of personal disaster scenarios such as air crashes or AIDS cannot be discounted, we can take some comfort in the fact that we are, as individuals, *statistically* unlikely to be affected. The acausal connection, though, could affect all of us and is in some ways the most frightening of all, because nothing would have changed, but everything would have changed. Pirsig calls this a 'Copernican inversion', after the way in which Copernicus's heliocentric (sun-centred) model of the solar system replaced the old geocentric (Earth-centred) version: the planets still continued to move in the same way, and life continued as normal, but the *way in which we saw* these familiar phenomena was completely changed. The willingness of people to accept acausality is shown by the growing popularity of religion, which is essentially acausal; and if acausal connections exist where do we stand?

It may have occurred to you that this last chapter reads very much like a horoscope — 'avoid air travel and making important decisions', for example — but that it is actually based on logically argued premises. If logic and causality alone can lead us into the realms of the unknown, I leave you with this question: can scientific thought perform a 'Copernican inversion' with respect to acausal relationships, and if it does will we be any nearer an understanding of the universe, or will we disappear into the Dark Ages of a new magic?

INDEX

Acausality 30, 119-20
AIDS 7, 91-9, 120
Airline crashes 77-88, 113
   Air India 747 78, 79, 87
   British Airtours 737 78, 80, 87-8
   China Airlines Flight 006 77-8, 79, 86-7
   Delta Airlines L-1011 78, 79, 87
   Galax Airlines Lockheed Electra 78, 79, 86
   Iberia 727 78, 79, 87
   Japan Airlines 747 78, 80-6
Alvarez, Luis and Walter 105
Apian, Peter 22
'Apollo' 103

Asteroid 103
Astrology 8, 9, 12-15
Astronomy 8, 9, 11, 40
Aurora Australis 35
Aurora Borealis 35

Bayeux Tapestry 41
Bhopal chemical plant disaster 7, 57, 58, 70, 71-5
Brahe, Tycho (Tyge) 15

Cassini, Giovanni 22
Clairaut 25
Cometary particles 100
'Cometary winters' 106
Cometology 11, 12
Comets

Bennett   28, 53, 66, 109
Delavan's   29, 49
Encke's   102
frequency of   28-9
Halley's   7, 12, 28, 29, 32,
    37, 38-43, 49, 55, 57,
    98, 100, 108, 109, 110,
    116, 118
Kohoutek   28, 49, 109
nature of   26-8
Tago-Sato-Kosaka   28,
    109
West   28, 109
Computers   113-14
'Copernican inversion'   120
Craters   104
    Arizona   104, 107
    Chubb   104
    Meteor   104
    Vredefort Ring   104

Dam failure   54
'Death star'   106
'Dirty snowball'   26, 34, 97,
    102, 104
Diseases, comet-borne   97-8,
    114-15, 119

Earthquakes   44-9, 111-13
    Mexico City   7, 44-9
    San Francisco   111
    Russia, Central   50, 108
'Earthquake-proofing'   46,
    47, 112
Einstein, Albert   17, 22
'Envelope'   27

Fires
    Australian bush   7, 51
    Bradford football stadium
        70, 75-6
Flamsteed   20, 22, 24
Floods   53, 54
'Flying sandbank'   26, 35
Football violence   59, 66-7
    Brussels   67

Galileo Galilei   15, 57
Gandhi, Indira, assassination
    of   7, 70
Gas explosion, Mexico City
    58, 70-1

Halley, Edmond   10, 12, 18,
    19-23, 24-5
Hijackings   67-9
    Achille Lauro   69, 108
    Jordanian 727   68
    Kuwaiti airliner   67-8
    TWA Flight 847   68-9
Hooke   18, 20, 21
Hoyle, Sir Fred   36, 39, 92,
    96-9, 106, 119
Hurricanes   53-5
    Elena   54
    Gloria   54

Jung, Carl Gustav 8, 29-30, 56

Kepler, Johannes 15, 16, 19, 20, 21, 22
Krakatoa 103

de Lalande, Joseph 25
Lepaute, Nicole 25
Lower, Sir William 16

'Mascons' 32
Meteorites 100-7
Mud-slide, Puerto Rico 108

'Nemesis' 106-7
Newton, Sir Isaac 16-19, 20, 21
Nostradamus 92
Notting Hill Carnival 59

Palitsch, George 25
Perihelion 12, 37, 113
Philadelphia 'Move' raid 59, 64-6
'Pills, anti-comet' 115
'Planet X' 106-7

'Radio storms' 35
Railway crashes 88-90, 113
    Argenton-sur-Creuse 88-9
    British 88
    French 98-90

    Portuguese 90
Riots
    Bristol 58, 60
    Brixton 58, 62
    football 59, 66-7
    Handsworth 58, 59-62
    Jamaican 58-9
    South Africa 58
    Tottenham 58, 63
    Toxteth 62
Royal Society 19, 20

San Andreas Fault 48, 111, 112
'Solar wind' 34, 35, 56, 97, 113, 114
Star of Bethlehem 38
'Sungrazer' 28
'Sun storms' 35
Synchronicity 8, 29-30, 37, 56, 88, 90, 109-10, 119

'Tail' 27, 34, 35, 97
Terrorism (see also Hijackings) 67-70
    Brighton bombing 69
Tidal waves 49, 107, 111
Tsunamis 49, 107, 111
Tunguska Incident 101-3, 105, 107
Typhoons 53-5
    Bay of Bengal 7, 56
    Kit 55

Nelson 55
Pat 55

'Wavicles' 30, 32
Weather
  freak 50-5
  patterns 50

Whipple, Fred 26
Wickramasinghe, Chandra 36, 39, 92, 96-9, 106, 119
Wren, Sir Christopher 21